普通高等学校"十四五"规划计算机类专业特色教材

计算机网络实践教程

（基于华为eNSP）

○ 主　审　卢　军
○ 主　编　陈　祥　朱　薏　黄进勇
○ 副主编　田　强　冯　晨　何文斌　卢心刚

华中科技大学出版社
http://press.hust.edu.cn
中国·武汉

内容简介

本书是介绍计算机网络主流技术的实践操作教材，覆盖了交换技术、路由技术、无线网络技术和网络安全技术等网络关键技术。在内容上以何文斌编著的《计算机网络》为依托，与教材中的相关理论相互印证，让读者能够更好地理解计算机网络的原理，并能获得一定的组网能力，为进一步学习打下坚实的基础。

本书可作为计算机网络实验教材和计算机网络工程教材，也可供计算机网络管理人员和工程技术人员在学习和研究计算机网络时参考。

图书在版编目(CIP)数据

计算机网络实践教程 / 陈祥，朱薏，黄进勇主编. —武汉：华中科技大学出版社，2023.8
ISBN 978-7-5680-9671-3

Ⅰ.①计… Ⅱ.①陈… ②朱… ③黄… Ⅲ.①计算机网络-教材 Ⅳ.①TP393

中国国家版本馆 CIP 数据核字(2023)第 154495 号

计算机网络实践教程 陈祥 朱薏 黄进勇 主编
Jisuanji Wangluo Shijian Jiaocheng

策划编辑：汪 粲
责任编辑：余 涛
封面设计：原色设计
责任监印：周治超

出版发行：华中科技大学出版社(中国·武汉) 电话：(027)81321913
　　　　　武汉市东湖新技术开发区华工科技园 邮编：430223
录　　排：华中科技大学惠友文印中心
印　　刷：武汉科源印刷设计有限公司
开　　本：787mm×1092mm　1/16
印　　张：20.75
字　　数：516 千字
版　　次：2023 年 8 月第 1 版第 1 次印刷
定　　价：56.00 元

本书若有印装质量问题，请向出版社营销中心调换
全国免费服务热线：400-6679-118　竭诚为您服务
版权所有　侵权必究

前　言

　　计算机网络作为计算机技术与通信技术密切结合的技术，是一门实践性非常强的课程。理论教学应与实践环节紧密结合，"计算机网络实验"和"计算机网络工程"等课程的教学对于网络人才的培养尤为重要。

　　本书是与何文斌编著的教材《计算机网络》配套的实验教材，也可以作为"计算机网络工程"课程的教材，详细介绍了在华为 eNSP 软件实验平台上完成组建以太网、交换机配置、路由器配置、VLAN 路由配置、广域网的配置、访问控制列表 ACL 配置、网络地址转换 NAT 配置、应用服务器配置、VPN 配置、防火墙配置、双机热备份配置、无线局域网配置等实验。

　　本书详细介绍了华为 eNSP 软件实验平台的功能和使用方法，全书分为 14 章，包括 43 个实验。每个实验都从实验原理、实验过程中使用的华为 VRP 命令和实验步骤三个方面进行深入讨论，不仅能使读者掌握用华为网络设备完成网络设计、配置与调试的过程和步骤，而且能使读者进一步理解实验所涉及的原理和技术。

　　华为 eNSP 软件的人机界面非常接近实际华为网络设备的配置过程，除了连接线缆等物理动作外，读者通过华为 eNSP 软件完成实验的过程与通过实际华为网络设备完成实验的过程几乎没有差别。华为 eNSP 软件通过与 Wireshark 结合，能够捕获经过主机、交换机、路由器和防火墙各个接口的报文，显示各个阶段应用层消息、传输层报文、IP 分组、封装 IP 分组的链路层帧的结构、内容和首部中每个字段的值都使读者可以直观了解 IP 分组的端到端传输过程，以及 IP 分组端到端传输过程中各层 PDU 的细节和变换过程。

　　"计算机网络"和"计算机网络工程"是实验性很强的课程，需要通过实际网络设计过程加深对教学内容的理解，以此培养学生分析问题、解决问题的能力。但实验又是一大难题，因为很少有学校可以提供包括设计、实施等各种类型网络的网络实验室，华为 eNSP 软件实验平台和本书很好地解决了这一难题。

　　作为与理论教材《计算机网络》配套的实验教材，本书和理论教材相得益彰，理论教材为读者提供了计算机网络的原理和设计方法，本书提供了在华为 eNSP 软件实验平台上运用理论教材提供的理论和方法设计、配置和调试各种类型网络的过程和步骤，读者用理论教材提供的网络设计原理和方法指导实验，反过来又通过实验加深对理论教材内容的理解，课堂教学和实验形成良性互动。

　　本书由湖北工程学院新技术学院的陈祥、黄进勇老师，以及湖北交通职业技术学院交通信息学院的朱薏老师担任主编，由湖北工程学院新技术学院的田强、冯晨、何文斌和卢心刚等老师任副主编。全书由湖北工程学院卢军教授任主审。

　　在本书编写过程中，作者参阅了大量同类书籍和网络资源，融合了许多自己的观点和见解，并力求做到深入浅出、通俗易懂，但由于作者水平和经验有限，不足之处在所难免，敬请同行专家批评指正。

编者
2023 年 6 月

目 录

第 1 章 eNSP 基本操作 ... 1
1.1 认识 eNSP ... 1
1.2 VRP 基本操作 ... 11
1.3 熟悉常用的 IP 相关命令 ... 16

第 2 章 常用网络命令 ... 22
2.1 ipconfig 命令 ... 22
2.2 ping 命令 ... 26
2.3 arp 命令 ... 32
2.4 netstat 命令 ... 35
2.5 tracert 命令 ... 40

第 3 章 组建以太网 ... 43
3.1 双绞线制作 ... 43
3.2 集线器和交换机组建以太网 ... 46

第 4 章 交换机配置 ... 54
4.1 交换机基本配置 ... 54
4.2 STP 配置 ... 58
4.3 VLAN 基本配置 ... 65
4.4 Eth-Trunk 链路聚合配置 ... 72

第 5 章 路由器配置 ... 83
5.1 静态路由配置 ... 83
5.2 RIP 动态路由配置 ... 91
5.3 OSPF 的动态路由配置 ... 100
5.4 BGP 邻居配置 ... 116

第 6 章 VLAN 路由配置 ... 126
6.1 利用单臂路由器实现 VLAN 间路由 ... 126
6.2 利用三层交换机实现 VLAN 间路由 ... 133

第 7 章 广域网的配置 ... 139
7.1 广域网接入配置 ... 139
7.2 帧中继配置 ... 145

计算机网络实践教程

第 8 章　访问控制列表 ACL 配置 151
8.1　基本 ACL 配置 151
8.2　高级 ACL 配置 155

第 9 章　网络地址转换 NAT 配置 161
9.1　静态 NAT 配置 161
9.2　动态 NAT 配置 168
9.3　NAPT 配置 173

第 10 章　应用服务器配置 180
10.1　FTP 服务器配置 180
10.2　Web 服务器配置 187
10.3　DNS 服务器配置 192
10.4　DHCP 服务器配置 197
10.5　Telnet 服务配置 208
10.6　SSH 配置 214

第 11 章　VPN 配置 222
11.1　IPSec 基本配置 222
11.2　Efficient VPN 配置 231
11.3　GRE VPN 协议基本配置 238
11.4　MPLS/LDP 基本配置 248
11.5　BGP/MPLS VPN 基本配置 259

第 12 章　防火墙配置 276

第 13 章　双机热备份配置 291
13.1　路由器热备份配置 291
13.2　交换机热备份配置 297
13.3　防火墙热备份配置 307

第 14 章　无线局域网配置 315

第 1 章 eNSP 基本操作

1.1 认识 eNSP

一、实验目的

1. 认识 eNSP
2. 掌握 eNSP 的安装方法
3. 了解 eNSP 的各种功能
4. 掌握运用 eNSP 搭建网络拓扑并进行实验的操作方法

二、实验内容及步骤

1. 安装

安装 eNSP 之前，需要先安装依赖组件，包括 Oracle VM VirtualBox、WinPcap、Wireshark 和 Vlcmediaplayer 等软件，需要安装软件列表如图 1.1 所示。

名称	修改日期	类型	大小
eNSP_Setup.exe	2022/7/15 9:00	应用程序	555,442 KB
VirtualBox-5.2.30-130521-Win.exe	2022/7/15 9:00	应用程序	113,184 KB
vlcmediaplayer3016.exe	2022/10/3 16:00	应用程序	41,761 KB
WinPcap_4_1_3 - 副本.zip	2022/8/4 23:34	WinRAR ZIP 压缩...	877 KB
Wireshark3.6.7 - 副本.rar	2022/8/4 23:44	WinRAR 压缩文件	134,938 KB

图 1.1 需要安装软件列表

2. 安装 eNSP

(1) 双击安装程序文件，打开安装向导。

(2) 在"选择安装语言"对话框中选择"中文（简体）"，单击"确定"按钮，如图 1.2 所示。

图 1.2　选择安装语言

（3）进入欢迎界面，单击"下一步"按钮，如图 1.3 所示。

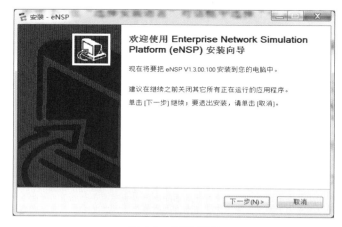

图 1.3　欢迎界面

（4）在"许可协议"页，选择"我愿意接受此协议"，单击"下一步"按钮，如图 1.4 所示。

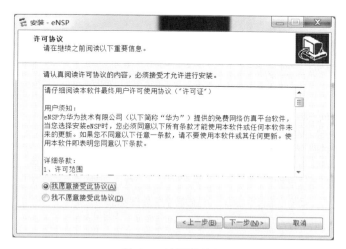

图 1.4　许可协议页

（5）设置安装的目录（整个目录路径都不能包含非英文字符），单击"下一步"按钮，如

图 1.5 所示。

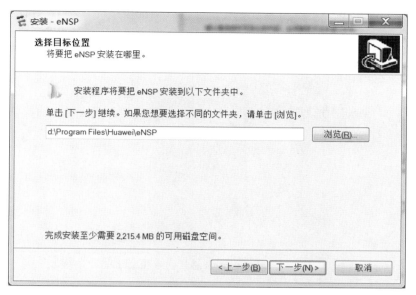

图 1.5 选择安装路径

(6) 设置 eNSP 程序快捷方式位置及名称,单击"下一步"按钮,如图 1.6 所示。

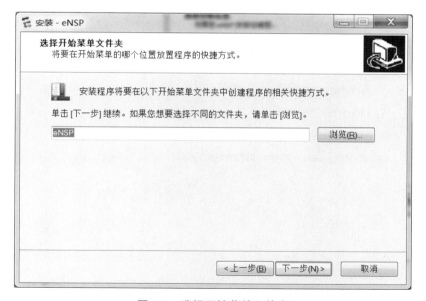

图 1.6 选择开始菜单文件夹

(7) 选择是否要在桌面创建快捷方式,单击"下一步"按钮,如图 1.7 所示。

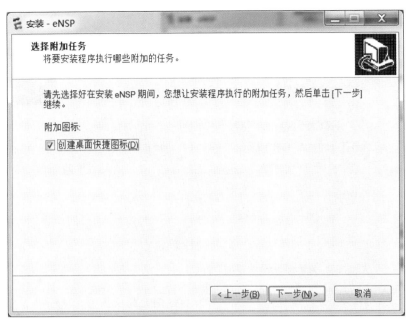

图 1.7 选择创建桌面快捷方式

(8) 选择安装其他程序，单击"下一步"按钮，如图 1.8 所示。

图 1.8 选择安装的组件

(9) 确认安装信息后，单击"安装"按钮开始安装，如图 1.9 所示。

第 1 章　eNSP 基本操作

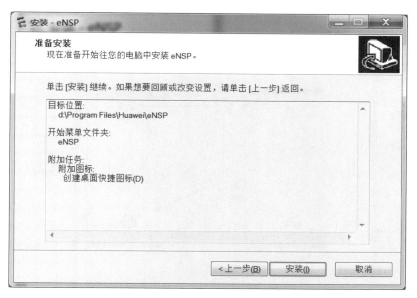

图 1.9　准备安装

（10）安装完成后，若不希望立刻打开程序，则可取消选中"运行 eNSP"复选框。单击"完成"按钮结束安装，如图 1.10 所示。

图 1.10　安装完成

3. 认识 eNSP 界面

启动 eNSP 模拟器软件，eNSP 主界面由五个区域组成，如图 1.11 所示。

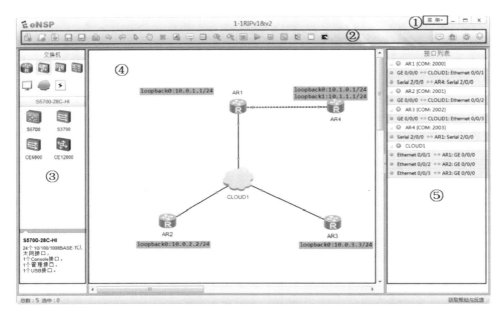

图 1.11 eNSP 主界面

区域①：主菜单，提供"文件""编辑""视图""工具"和"帮助"等菜单项；
区域②：工具栏，提供常用的工具，如新建/编辑拓扑、启动/停止设备等工具；
区域③：网络设备区，提供网络设备和设备连线等；
区域④：工作区，在此区域可以灵活创建和编辑网络拓扑；
区域⑤：设备接口列表，显示设备间连接的接口，指示灯显示接口运行状态。

4. 注册设备

eNSP 安装完成后，需要注册设备，单击 eNSP 主界面右上角的菜单，依次选择"工具"→"注册设备"选项，如图 1.12 所示。

图 1.12 "注册设备"选项

在"注册"对话框中勾选"所有类型的设备"，单击"注册"按钮。注册时会提示注册前先删除设备，注册功能会自动删除相关设备，然后自动注册成功，如图 1.13 所示。

第 1 章　eNSP 基本操作

图 1.13　注册设备

注册成功后，打开 Oracle VM VirtualBox 软件，查看 VirtualBox 中是否存在相关虚拟机，如图 1.14 所示。

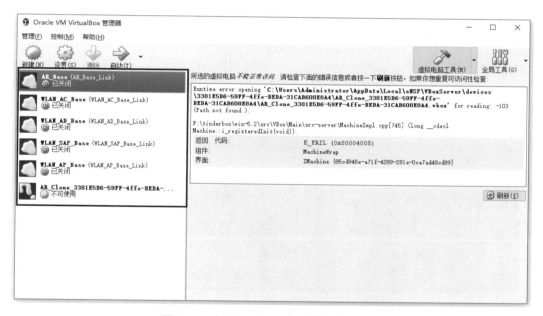

图 1.14　在 VirtualBox 中查看相关虚拟机

5．网络设备配置

在 eNSP 中，可以利用图形化界面灵活搭建需要的拓扑组网图，其步骤如下。

1) 选择设备

主界面左侧为可供选择的网络设备区，将需要的设备直接拖至工作区。每台设备带有默认名称，通过单击可以对其进行修改。还可以使用工具栏中的文本按钮和调色板按钮在拓扑中任意位置添加描述或图形标识，如图 1.15 所示。

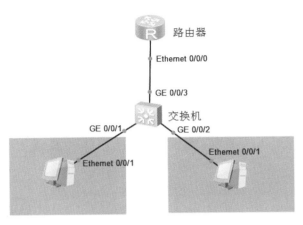

图 1.15 搭建拓扑

2) 配置设备

在拓扑中的设备图标上单击鼠标右键,在弹出的快捷菜单中选择"设置"命令,打开设备接口配置界面。

在"视图"选项卡中,可以查看设备面板及可供使用的接口卡,如图 1.16 所示。如需为设备增加接口卡,可在"eNSP 支持的接口卡"区域选择合适的接口卡,直接拖至上方的设备面板上相应空槽位即可;如需删除某个接口卡,直接将设备面板上的接口卡拖回"eNSP 支持的接口卡"区域即可。注意,只有在设备电源关闭的情况下才能进行增加或删除接口卡的操作。

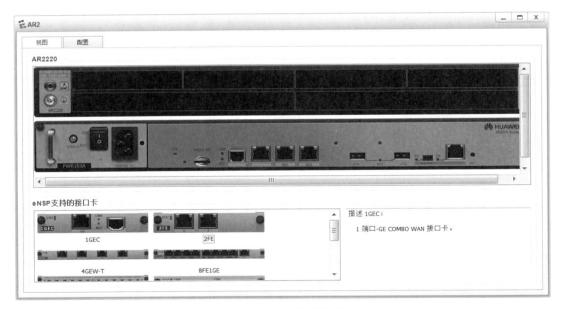

图 1.16 设备配置界面

在"配置"选项卡中,可以设置设备的串口号,串口号范围为 2000~65535,默认情况下从起始数字 2000 开始使用。可以自行更改串口号并单击"应用"按钮,如图 1.17 所示。

第 1 章　eNSP 基本操作

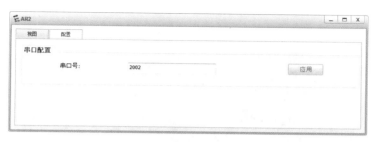

图 1.17　配置设备

在模拟 PC 图标上单击鼠标右键（或双击图标），在弹出的快捷菜单中选择"设置"命令，打开设置的对话框。在"基础配置"选项卡中配置设备的基础参数，如 IP 地址、子网掩码和 MAC 地址等，如图 1.18 所示。

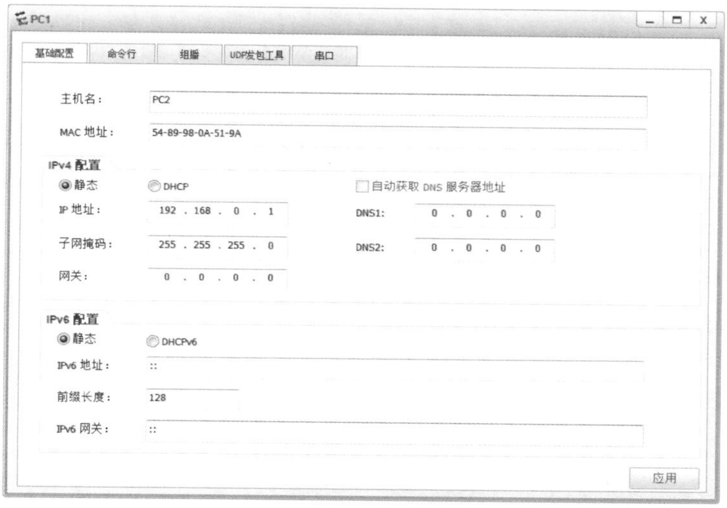

图 1.18　PC 配置界面

在"命令行"选项页可以输入 ping 命令，测试连通性，如图 1.19 所示。

3) 设备连接

根据设备接口的不同可以灵活选择线缆的类型。当线缆仅一端连接了设备，而此时希望取消连接时，在工作区单击鼠标右键或者按<Esc>键即可。选择"Auto"可识别接口卡类型，自动选择相应线缆。常见的如"Copper"为双绞线，"Serial"为串口线。

4) 配置导入

在设备未启动的状态下，在设备上单击鼠标右键，在快捷菜单中选择"导入设备配置"命令，可以选择设备配置文件（.cfg 或.zip）并导入设备中。

5) 设备启动

选中需要启动的设备后，可以通过单击工具栏中的"启动设备"按钮或者选择该设备的右键菜单的"启动"命令来启动设备。若未选择任何设备，则默认启动所有设备。启动后，双击设备图标，通过弹出的 CLI 命令行界面进行配置。

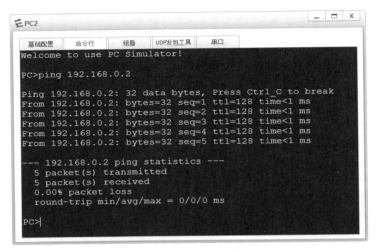

图 1.19　PC 命令行

6) 设备和拓扑保存

完成配置后可以单击工具栏中的"保存"按钮来保存拓扑，并导出设备的配置文件。在设备上单击鼠标右键，在快捷菜单中选择"导出设备配置"命令，输入设备配置文件的文件名，并将设备配置信息导出为.cfg 文件。

6．扩展功能介绍

1) 样例加载

在工具栏中单击"打开"按钮，可弹出 eNSP 附带的验证各种网络协议特性的典型实验案例。打开样例结果显示清晰的拓扑组网，如图 1.20 所示。

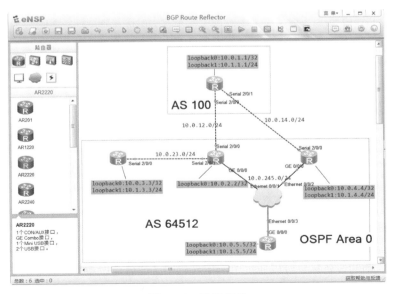

图 1.20　样例拓扑

第 1 章 eNSP 基本操作

每个样例都包括了具体的实验配置，如同一个配置好的真实的网络环境。当运行拓扑中所有的设备后，设备会自动加载配置，用户可以在这套模拟环境中学习和验证理论知识。

2) 数据抓包

设备运行时，在设备接口处单击鼠标右键，在快捷菜单中可选择"数据抓包"命令。通过抓包，用户可直观感受到数据包的流动，更深刻地理解网络体系结构。

3) 支持与真实 PC 进行桥接

通过虚拟设备接口与真实网卡的绑定，实现虚拟设备与真实环境的对接。

4) 支持使用第三方软件登录 eNSP 模拟设备

在 eNSP 软件的接口视图中，设备名称后面显示一个串口号，该端口号即使用第三方工具时需要设置的串口号。

1.2 VRP 基本操作

VRP（versatile routing platform，通用路由平台）是华为公司数据通信产品通用的网络操作系统平台，拥有一致的网络界面、用户界面和管理界面。在 VRP 中，用户通过命令行对设备下发各种命令来实现对设备的配置与维护。用户登录到路由器出现命令行提示符后，即进入命令行接口 CLI（command line interface）。命令行接口是用户与路由器进行交互的工具。

一、实验目的

1. 熟悉 VRP 的基本操作
2. 掌握命令行视图的切换
3. 掌握命令行帮助和快捷键的使用
4. 掌握修改设备名称和设置时钟的方法
5. 掌握设置标题信息的方法
6. 掌握查看路由器基本信息的方法

二、实验内容

本实验模拟用户首次使用 VRP 操作系统的过程。在登录路由器后使用命令行来配置设备，进行命令行视图的切换、命令行帮助和快捷键的使用，并完成设备的基本配置，包括修改路由器名称、配置路由器时钟、设置标题文本以及使用命令行查看路由器基本信息等。

三、实验拓扑

VRP 基本操作的拓扑如图 1.21 所示。

图 1.21 VRP 基本操作拓扑

四、实验步骤

1. 命令视图切换

启动设备，登录设备成功后即进入用户视图，如图 1.22 所示。

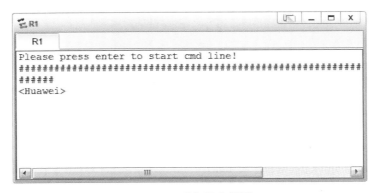

图 1.22 设备用户视图

在用户视图下只能使用参数和监控级命令，如使用 display version 命令显示系统软件版本及硬件等信息。

```
<Huawei>display version
Huawei Versatile Routing Platform Software
VRP (R) software, Version 5.130 (AR200 V200R003C00)
Copyright (C) 2011-2012 HUAWEI TECH CO., LTD
Huawei AR201 Router uptime is 0 week, 0 day, 0 hour, 2 minutes
```

结果显示，VRP 操作系统的版本为 5.130，设备型号为 V200R003C00，启动时间为 2 分钟等信息。

在用户视图下使用 system-view 命令可以切换到系统视图。在系统视图下可以配置接口、协议等，使用 quit 命令又可以切换回用户视图。

```
<Huawei>system-view
Enter system view, return user view with Ctrl+Z.
[Huawei]quit
<Huawei>
```

在系统视图下使用相应命令可进入其他视图，如使用 interface 命令进入接口视图。在接口视图下可以使用 ip address 命令配置接口 IP 地址、子网掩码。

为路由器的 GE 0/0/0 接口配置 IP 地址时可以使用子网掩码长度，也可以使用完整的子网掩码，如子网掩码为 255.255.255.0，可以使用 24 替代。

```
[Huawei]interface Ethernet 0/0/0
[Huawei-Ethernet0/0/0]ip address 192.168.1.1 24
```

配置完成后，可以使用 return 命令（或使用快捷键<Ctrl+Z>）直接退回到用户视图。

2．命令行帮助

如果用户忘记命令的参数或关键字，可使用命令行在线帮助。命令行在线帮助分为完全帮助和部分帮助。

1）完全帮助

在任意命令视图下，输入"?"获取该命令视图下所有的命令及其简单的描述。如在系统视图下，输入"?"获取帮助信息。

```
[Huawei]?
System view commands:
  aaa                       <Group> aaa command group
  aaa-authen-bypass         Set remote authentication bypass
  aaa-author-bypass         Set remote authorization bypass
  aaa-author-cmd-bypass     Set remote command authorization bypass
  access-user               User access
  acl                       Specify ACL configuration information
  ……
```

也可以输入一个命令，后接以空格分隔的"?"，列出全部关键字或参数及其简单描述。如在系统视图下，列出 interface 命令参数及其简单描述。

```
[Huawei]interface ?
  Bridge-if     Bridge-if interface
  Dialer        Dialer interface
  Eth-Trunk     Ethernet-Trunk interface
  Ethernet      Ethernet interface
  ……
```

2）部分帮助

输入一个字符串，其后紧接"?"，列出以该字符串开头的所有关键字。例如，在系统视图下列出以"re"字符串开头的所有命令及其简单描述。

```
[Huawei]re?
 refresh  <Group> refresh command group
 remove   Remove
 reset    <Group> reset command group
 return   Enter the privileged mode
```

3. 快捷键使用

命令行接口提供了基本的命令编辑功能，支持多行编辑，每条命令的最大长度为256个字符。各快捷键详细描述如表1.1所示。

表 1.1　快捷键及其功能

快捷键	功能
\<Backspace\>	删除光标位置的前一个字符
\<←\>或\<Ctrl+B\>	光标向左移动一个字符位置
\<→\>或\<Ctrl+F\>	光标向右移动一个字符位置
\<Delete\>	删除光标位置字符
\<↑\>或\<↓\>	显示历史命令
\<Tab\>	输入不完整命令，按该键可以补全命令

4. 修改设备名称

用户可以为每个设备设置特定的名称，以便于管理和识别设备，如使用 sysname 命令将当前设备默认名称改为 R1。

```
[Huawei]sysname R1
[R1]
```

5. 设置路由器时钟

为了保证设备有准确的时钟信号，用户需要准确设置设备的系统时钟。如在用户视图下，使用 clock datetime 命令修改系统日期和时间为2022年9月16日10时18分18秒，使用 clock timezone 命令，设置所在的时区为北京。

```
<R1>clock datetime 10:18:18 2022-09-16
<R1>clock timezone BJ add 08:00:00
```

6. 设置标题信息

如果需要对登录路由器的用户提供警示或说明信息，可以设置登录时或登录成功后的标题信息。使用 header login information 命令，可设置登录时的标题文本为"hello"，使用 header shell information 命令，可设置登录成功后的标题文本信息为"Welcome to Huawei certification lab"。

```
[R1]header login information "hello"
[R1]header shell information "Welcome to Huawei certification lab"
```

配置完成后，使用 quit 命令尝试退出路由器命令重新登录，可观察到欢迎信息。

```
[R1]quit
<R1>quit
Configuration console exit, please press any key to log on
Welcome to Huawei certification lab
<R1>
```

7. 查看路由器基本信息

使用 display 系列命令可查看路由器基本信息或运行状态。
(1) 查看设备版本信息。
使用 display version 命令查看路由器的版本信息。

```
<R1>display version
Huawei Versatile Routing Platform Software
VRP (R) software, Version 5.130 (AR200 V200R003C00)
Copyright (C) 2011-2012 HUAWEI TECH CO., LTD
......
```

(2) 查看设备当前配置。
使用 display current-configuration 命令查看路由器当前配置，可以查看到设备名称、登录标题信息、时区等信息。

```
<R1>display current-configuration
[V200R003C00]
#
 sysname R1
 header shell information "Welcome to Huawei certification lab"
 header login information "hello"
#
 snmp-agent local-engineid 800007DB03000000000000
```

```
 snmp-agent
 #
 clock timezone BJ add 08:00:00
 #
 ......
 Return
```

(3) 使用 display interface Ethernet 0/0/0 命令查看路由器 Ethernet 0/0/0 接口的状态。结果显示该接口的物理状态、接口 IP 地址以及其他统计信息。

```
[R1]display interface Ethernet 0/0/0
Ethernet0/0/0 current state : UP
Line protocol current state : UP
Description:HUAWEI, AR Series, Ethernet0/0/0 Interface
Route Port, The Maximum Transmit Unit is 1500
Internet Address is 192.168.1.1/24
IP Sending Frames'Format is PKTFMT_ETHNT 2,Hardware address is
00e0-fc33-3bcf
 ......
```

(4) 使用 display this 命令查看当前视图下生效的运行配置参数。结果显示该视图下生效的 IP 地址、RIP 版本号等信息。

```
[R1-GigabitEthernet0/0/0]display this
[V200R003C00]
#
interface GigabitEthernet0/0/0
 ip address 10.0.123.1 255.255.255.0
 rip version 1
#
return
[R1-GigabitEthernet0/0/0]
```

1.3 熟悉常用的 IP 相关命令

华为设备支持多种配置方式，包括 Web 界面管理等。作为一名网络工程师，必须熟悉使用

第 1 章　eNSP 基本操作

命令行的方式进行设备管理。在工作中，对路由器和交换机最常用的操作命令是 IP 相关命令，如配置主机名、IP 地址、测试 IP 数据包连通性等。这些命令是基本的配置和测试命令。

一、实验目的

1. 掌握路由器命名的方法
2. 掌握配置路由器 IP 地址方法
3. 掌握测试 IP 地址连通性的方法
4. 掌握查看设备配置的方法
5. 掌握抓包的方法

二、实验内容

本实验模拟简单的企业网络场景，某公司购买了新的路由器和交换机。交换机 SW1 连接客服部 PC1，SW2 连接市场部 PC2，路由器 R1 连接 SW1 和 SW2 两台交换机。网络管理员需要首先熟悉设备的使用，包括基础的 IP 配置和查看命令。

三、实验拓扑

本实验的拓扑如图 1.23 所示。

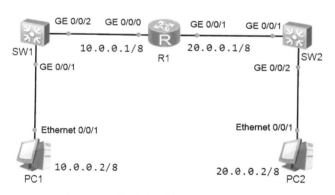

图 1.23　熟悉常用的 IP 相关命令拓扑

四、实验地址分配

实验地址分配如表 1.2 所示。

表 1.2　实验地址分配

设备	接口	IP 地址	子网掩码	默认网关
PC1	Ethernet 0/0/1	10.0.0.2	255.0.0.0	10.0.0.1
PC2	Ethernet 0/0/1	20.0.0.2	255.0.0.0	20.0.0.1

续表

设备	接口	IP 地址	子网掩码	默认网关
R1（AR2220）	GE 0/0/0	10.0.0.1	255.0.0.0	N/A
	GE 0/0/1	20.0.0.1	255.0.0.0	N/A

五、实验步骤

1. 基本配置

1) 计算机的配置

根据实验地址分配，使用图形化界面配置 PC 的 IP 地址，以 PC1 为例，如图 1.24 所示。

图 1.24　PC1 配置界面

PC1 设置完成后，继续设置 PC2，如图 1.25 所示。

图 1.25　PC2 配置界面

第 1 章　eNSP 基本操作

2）路由器的配置

在路由器 R1 上使用 system-view 命令从用户视图切换到系统视图，使用 sysname 命令命名路由器为 R1，在完成设备的配置后，可以使用 save 命令对完成的配置进行保存。

```
<Huawei>system-view        //由用户视图切换到系统视图
Enter system view, return user view with Ctrl+Z.
[Huawei]sysname R1         //修改设备主机名为R1
[R1]quit
<R1>save                   //保存当前配置
The current configuration will be written to the device.
Are you sure to continue? (y/n)[n]:y
It will take several minutes to save configuration file, please wait........
Configuration file had been saved successfully
Note: The configuration file will take effect after being activated
```

2. 配置路由器接口 IP 地址

在路由器 R1 上的系统视图中，使用 interface 命令切换到某一个接口，对接口进行配置。使用 ip address 命令为接口配置 IP 地址，配置完成后，使用 display ip interface brief 命令查看接口与 IP 地址的摘要信息。

```
[R1]interface GigabitEthernet 0/0/0    //切换到GE 0/0/0端口
[R1-GigabitEthernet0/0/0]ip address 10.0.0.1 255.0.0.0  //配置IP地址
[R1-GigabitEthernet0/0/0]quit
[R1]interface GigabitEthernet 0/0/1    //切换到GE 0/0/0端口
[R1-GigabitEthernet0/0/1]ip address 20.0.0.1 255.0.0.0  //配置IP地址
[R1-GigabitEthernet0/0/1]quit
[R1]display ip interface brief  //查看接口与IP摘要信息
Interface                IP Address/Mask      Physical    Protocol
GigabitEthernet0/0/0     10.0.0.1/8           up          up
GigabitEthernet0/0/1     20.0.0.1/8           up          up
NULL0                    unassigned           down        down
```

结果显示，路由器配置了 GE 0/0/0 和 GE 0/0/1 接口的 IP 地址，接口的物理状态为激活状态。

3. 查看路由器配置信息

1）查看路由表

经过以上步骤的配置，路由器接口的 IP 地址已经配置完成，可以使用 display ip routing-table 命令查看 IPv4 路由表的信息。

```
<R1>display ip routing-table
Route Flags: R - relay, D - download to fib
------------------------------------------------------------
Routing Tables: Public
        Destinations : 10        Routes : 10
Destination/Mask   Proto   Pre  Cost   Flags  NextHop    Interface
      10.0.0.0/8   Direct  0    0      D      10.0.0.1   GigabitEthernet 0/0/0
      ......
      20.0.0.0/8   Direct  0    0      D      20.0.0.1   GigabitEthernet 0/0/1
      ......
```

结果显示，路由器 R1 在 GE 0/0/0 接口上直连了一个 10.0.0.0/8 的网段，在 GE 0/0/1 接口上直连了一个 20.0.0.0/8 的网段。

2) 测试连通性

在 R1 上使用 ping 命令测试与 PC1 的连通性。

```
<R1>ping 10.0.0.2
 ping 10.0.0.2: 56  data bytes, press CTRL_C to break
  Reply from 10.0.0.2: bytes=56 Sequence=1 ttl=128 time=130 ms
  ......
  Reply from 10.0.0.2: bytes=56 Sequence=5 ttl=128 time=50 ms
--- 10.0.0.2 ping statistics ---
  5 packet(s) transmitted
  5 packet(s) received
  0.00% packet loss
  round-trip min/avg/max = 30/58/130 ms
```

结果显示，R1 与 PC1 通信正常。同样可以测试 R1 与 PC2、PC1 与 PC2 之间的连通性。

4. 使用抓包工具

以抓去 R1 上 GE 0/0/0 接口的数据包为例，在 R1 与 SW1 的直连链路上，在接口 GE 0/0/0 上单击鼠标右键，在弹出的快捷菜单中选择"开始抓包"命令，如图 1.26 所示。也可以通过接口列表区域直接选择接口，然后右键选择"开始数据抓包"命令。

图 1.26 右键抓包示意图

这时会显示出解析数据包的结果，如图 1.27 所示。双击数据包可以查看详细数据包的内容，例如，图 1.27 中可以展开 IEEE 802.3 Ethernet 的数据帧，显示目的 MAC 地址和源 MAC 地址等。如不需要继续抓包，可在接口的快捷菜单中选择"停止抓包"命令。

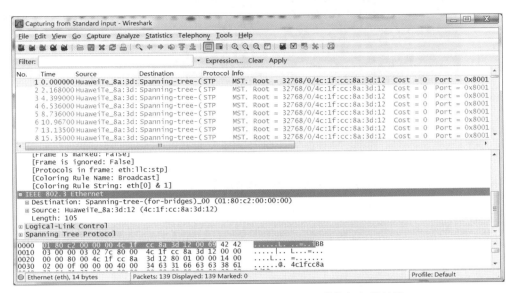

图 1.27　解析通过 R1 的 GE 0/0/0 接口的数据包的结果

第 2 章 常用网络命令

本章以 Windows 系统下的命令为例对常用网络命令进行操作。

2.1 ipconfig 命令

一、原理概述

ipconfig 命令用于显示、更新和释放网络地址设置，包括 IP 地址、子网掩码、默认网关、DNS 服务器设置等。

Windows 系统用户通过执行 "**win+R**" 组合键，在弹出的窗口中输入 "**cmd**" 来打开命令行程序。在提示符后，按如下格式输入：

```
ipconfig [/allcompartments][/?|/all|/renew[adapter]|/release
        [adapter]|/renew6[adapter]|/release6 [adapter]|
        /flushdns|/displaydns | /registerdns| /showclassid
        adapter |/setclassid adapter [classid] |/showclassid6
        adapter|/setclassid6 adapter [classid]]
```

ipconfig 命令的常见参数及其含义如下：

```
/?                  显示帮助消息
/all                显示完整配置信息
/renew              新指定适配器的IPv4地址
/renew6             更新指定适配器的IPv6地址
/release            释放指定适配器的IPv4地址
/flushdns           清除DNS解析程序缓存的内容
```

第 2 章 常用网络命令

```
/displaydns          显示DNS解析程序缓存的内容
```

二、实验目的

1. 掌握 ipconfig 命令的基本功能
2. 掌握 ipconfig 命令的几个常用参数使用方法

三、实验内容

在接入局域网或 Internet 的计算机上使用 ipconfig 命令查看基本信息、查看详细信息、更新自动获取的 IP 地址、释放自动获取的 IP 地址、查看 DNS 缓存和清除 DNS 缓存。

四、实验步骤

1. 查看 TCP/IP 的基本信息

Windows 系统用户通过执行"**win+R**"组合键，在弹出的窗口中输入"**cmd**"来打开命令行程序。在提示符后输入"**ipconfig**"，结果如下：

```
C:\Users\Administrator>ipconfig
以太网适配器 以太网：
   连接特定的 DNS 后缀 . . . . . . . . : lan
   本地链接 IPv6 地址. . . . . . . . : fe80::82c7:aa8d:b250:4af6%14
   IPv4 地址 . . . . . . . . . . . . : 192.168.110.8
   子网掩码  . . . . . . . . . . . . : 255.255.255.0
   默认网关. . . . . . . . . . . . . : 192.168.110.1
```

2. 查看 TCP/IP 的详细信息

Windows 系统用户通过执行"**win+R**"组合键，在弹出的窗口中输入"**cmd**"来打开命令行程序。在提示符后输入"**ipconfig /all**"，结果如下：

```
C:\Users\Administrator>ipconfig/all
以太网适配器 以太网：
   连接特定的 DNS 后缀 . . . . . . . . : lan
   描述. . . . . . . . . . . . . . . : Realtek PCIe GbE Family Controller
   物理地址. . . . . . . . . . . . . : 1C-69-7A-B9-04-DC
   DHCP 已启用 . . . . . . . . . . . : 是
   自动配置已启用. . . . . . . . . . : 是
   本地链接 IPv6 地址. . . . . . . . : fe80::82c7:aa8d:b250:4af6%14(首选)
   IPv4 地址 . . . . . . . . . . . . : 192.168.110.8(首选)
```

```
    子网掩码 . . . . . . . . . . . . . . . : 255.255.255.0
    获得租约的时间 . . . . . . . . . . : 2023年6月14日 3:47:21
    租约过期的时间 . . . . . . . . . . : 2023年6月27日 11:54:22
    默认网关. . . . . . . . . . . . . . . : 192.168.110.1
    DHCP 服务器 . . . . . . . . . . . . : 192.168.110.1
    DHCPv6 IAID . . . . . . . . . . . . : 115392003
    DHCPv6 客户端 DUID . . . . . . . . :
    00-01-00-01-2A-A0-8D-8D-1C-69-7A-B9-04-DC
    DNS 服务器 . . . . . . . . . . . . . : 192.168.110.1
    TCPIP 上的 NetBIOS . . . . . . . . : 已启用
```

3. 释放自动获取的 IP 地址

Windows 系统用户通过执行 "win+R" 组合键，在弹出的窗口中输入 "cmd" 来打开命令行程序。在提示符后输入 "ipconfig /release"，结果如下：

```
C:\Users\Administrator>ipconfig/release
以太网适配器 以太网:
    连接特定的 DNS 后缀 . . . . . . . :
    本地链接 IPv6 地址. . . . . . . . : fe80::82c7:aa8d:b250:4af6%14
    默认网关. . . . . . . . . . . . . :
```

4. 更新自动获取的 IP 地址

Windows 系统用户通过执行 "win+R" 组合键，在弹出的窗口中输入 "cmd" 来打开命令行程序。在提示符后输入 "ipconfig /renew"，结果如下：

```
C:\Users\Administrator>ipconfig/renew
以太网适配器 以太网:
    连接特定的 DNS 后缀 . . . . . . . : lan
    本地链接 IPv6 地址. . . . . . . . : fe80::82c7:aa8d:b250:4af6%14
    IPv4 地址 . . . . . . . . . . . . : 192.168.110.8
    子网掩码 . . . . . . . . . . . . . : 255.255.255.0
    默认网关. . . . . . . . . . . . . : 192.168.110.1
```

5. 清除 DNS 解析程序缓存的内容

Windows 系统用户通过执行 "win+R" 组合键，在弹出的窗口中输入 "cmd" 来打开命令行程序。在提示符后输入 "ipconfig /flushdns"，结果如下：

```
C:\Users\Administrator>ipconfig /flushdns
Windows IP 配置
已成功刷新 DNS 解析缓存。
```

6. 查看 DNS 解析程序缓存的内容

先访问 www.baidu.com 网站。

Windows 系统用户通过 "**win+R**" 组合键，在弹出的窗口中输入 "**cmd**" 来打开命令行程序。在提示符后输入 "**ipconfig /displaydns**"，结果如下：

```
C:\Users\Administrator>ipconfig /displaydns
Windows IP 配置
    ......
    www.baidu.com
    ----------------------------------------
    记录名称 . . . . . . . : www.baidu.com
    记录类型 . . . . . . . : 5
    生存时间 . . . . . . . : 167
    数据长度 . . . . . . . : 8
    部分 . . . . . . . . . : 答案
    CNAME 记录 . . . . . : www.a.shifen.com

    记录名称 . . . . . . . : www.a.shifen.com
    记录类型 . . . . . . . : 1
    生存时间 . . . . . . . : 167
    数据长度 . . . . . . . : 4
    部分 . . . . . . . . . : 答案
    A (主机)记录 . . . . : 112.80.248.76

    记录名称 . . . . . . . : www.a.shifen.com
    记录类型 . . . . . . . : 1
    生存时间 . . . . . . . : 167
    数据长度 . . . . . . . : 4
    部分 . . . . . . . . . : 答案
    A (主机)记录 . . . . : 112.80.248.75
    ......
```

通过访问 www.baidu.com，在 DNS 缓存中出现 www.baidu.com 的域名缓存记录。记录中显示域名的别名为 www.a.shifen.com，域名对应的 IP 地址有多个。

计算机网络实践教程

2.2　ping 命令

一、原理概述

ping 命令是最常用的命令，特别是在组网中。ping 命令基于 ICMP，在源站执行，向目的站发送 ICMP 回送请求报文，目的站在收到报文后向源站返回 ICMP 回送应答报文，源站把返回的结果信息显示出来。

ping 命令用来测试站点之间是否可达，若可达，则可进一步判断双方的通信质量，包括稳定性等。

需要注意的是，有些主机为了防止通过 ping 探测，利用防火墙设置禁止 ICMP 报文，这样就不能通过 ping 命令确定该主机是否处于连通状态。

Windows 系统用户通过执行"win+R"组合键，在弹出的窗口中输入"cmd"来打开命令行程序。在提示符后，按如下格式输入：

```
ping [-t][-a][-n count][-l size][-f][-i TTL][-v TOS][-r count] [-s count]
[[-j host-list] | [-k host-list]] [-w timeout] [-R] [-S srcaddr]
[-c compartment][-p][-4] [-6] target_names
```

ping命令的常见参数及其含义如下：

```
-t          ping指定的主机，直到停止。若要查看统计信息并继续操作ping，则按下
            Ctrl+Break组合键；若要停止ping，则按下Ctrl+C组合键
-a          将地址解析成主机名
-n count    要发送的回显请求数
-l size     发送缓冲区大小
-i TTL      生存时间
-r count    记录计数跃点的路由（仅适用于IPv4）
-4          强制使用IPv4
-6          强制使用IPv6
```

二、实验目的

1. 掌握 ping 命令的基本功能
2. 掌握 ping 命令的几个常用参数使用方法
3. 掌握网络故障检测的一般步骤

三、实验内容

在接入 Internet 的计算机上使用 ping 命令检测网络的故障。

四、实验步骤

1. 不带参数的 ping 命令

Windows 系统用户通过执行"**win+R**"组合键，在弹出的窗口中输入"**cmd**"来打开命令行程序。在提示符后输入"**ping 127.0.0.1**"，结果如下：

```
C:\Users\Administrator>ping 127.0.0.1
正在 ping 127.0.0.1 具有 32 字节的数据:
来自 127.0.0.1 的回复: 字节=32 时间<1ms TTL=64
来自 127.0.0.1 的回复: 字节=32 时间<1ms TTL=64
来自 127.0.0.1 的回复: 字节=32 时间<1ms TTL=64
来自 127.0.0.1 的回复: 字节=32 时间<1ms TTL=64
127.0.0.1 的 ping 统计信息:
    数据包: 已发送 = 4, 已接收 = 4, 丢失 = 0 (0% 丢失),
往返行程的估计时间(以毫秒为单位):
    最短 = 0ms, 最长 = 0ms, 平均 = 0ms
```

本机向目标主机默认发送 4 个 32 字节的数据包，目标主机返回 4 个 32 字节的数据包，时间均小于 1 ms。结果显示，本机与目标主机之间是可达的。

2. ping -t

Windows 系统用户通过执行"**win+R**"组合键，在弹出的窗口中输入"**cmd**"来打开命令行程序。在提示符后输入"**ping –t 127.0.0.1**"，结果如下：

```
C:\Users\Administrator>ping -t 127.0.0.1
正在 ping 127.0.0.1 具有 32 字节的数据:
来自 127.0.0.1 的回复: 字节=32 时间<1ms TTL=64
来自 127.0.0.1 的回复: 字节=32 时间<1ms TTL=64
来自 127.0.0.1 的回复: 字节=32 时间<1ms TTL=64
来自 127.0.0.1 的回复: 字节=32 时间<1ms TTL=64
来自 127.0.0.1 的回复: 字节=32 时间<1ms TTL=64
来自 127.0.0.1 的回复: 字节=32 时间<1ms TTL=64
来自 127.0.0.1 的回复: 字节=32 时间<1ms TTL=64
来自 127.0.0.1 的回复: 字节=32 时间<1ms TTL=64
来自 127.0.0.1 的回复: 字节=32 时间<1ms TTL=64
127.0.0.1 的 ping 统计信息:
    数据包: 已发送 = 9, 已接收 = 9, 丢失 = 0 (0% 丢失),
```

往返行程的估计时间(以毫秒为单位):
    最短 = 0ms, 最长 = 0ms, 平均 = 0ms
Control-C
^C
C:\Users\Administrator>
```

本机向目标主机不间断发送 32 字节的数据包,在发送 9 个时,在键盘上输入"Ctrl+C"强行终止。结果显示,当网络出现故障时,可以使用该命令检测检修过程中显示网络连通性。

3. ping -n

Windows 系统用户通过执行"win+R"组合键,在弹出的窗口中输入"cmd"来打开命令行程序。在提示符后输入"ping –n 6 127.0.0.1",结果如下:

```
C:\Users\Administrator>ping -n 6 127.0.0.1
正在 ping 127.0.0.1 具有 32 字节的数据:
来自 127.0.0.1 的回复: 字节=32 时间<1ms TTL=64
来自 127.0.0.1 的回复: 字节=32 时间<1ms TTL=64
来自 127.0.0.1 的回复: 字节=32 时间<1ms TTL=64
来自 127.0.0.1 的回复: 字节=32 时间<1ms TTL=64
来自 127.0.0.1 的回复: 字节=32 时间<1ms TTL=64
来自 127.0.0.1 的回复: 字节=32 时间<1ms TTL=64
127.0.0.1 的 ping 统计信息:
    数据包: 已发送 = 6, 已接收 = 6, 丢失 = 0 (0% 丢失),
往返行程的估计时间(以毫秒为单位):
    最短 = 0ms, 最长 = 0ms, 平均 = 0ms
```

本机向目标主机发送 6 个 32 字节的数据包,目标主机返回 6 个 32 字节的数据包。结果显示,发送和返回数据包的个数由默认的 4 个变成 6 个。

4. ping -l

Windows 系统用户通过执行"win+R"组合键,在弹出的窗口中输入"cmd"来打开命令行程序。在提示符后输入"ping –l 64 127.0.0.1",结果如下:

```
C:\Users\Administrator>ping l 64 127.0.0.1
正在 ping 127.0.0.1 具有 64 字节的数据:
来自 127.0.0.1 的回复: 字节=64 时间<1ms TTL=64
来自 127.0.0.1 的回复: 字节=64 时间<1ms TTL=64
来自 127.0.0.1 的回复: 字节=64 时间<1ms TTL=64
来自 127.0.0.1 的回复: 字节=64 时间<1ms TTL=64
127.0.0.1 的 ping 统计信息:
    数据包: 已发送 = 4, 已接收 = 4, 丢失 = 0 (0% 丢失),
```
```

往返行程的估计时间(以毫秒为单位)：
    最短 = 0ms, 最长 = 0ms, 平均 = 0ms

本机向目标主机发送 4 个 64 字节的数据包，目标主机返回 4 个 64 字节的数据包。结果显示，发送和返回数据包的大小由默认的 32 字节变成 64 字节。

5. ping -a

Windows 系统用户通过执行"win+R"组合键，在弹出的窗口中输入"cmd"来打开命令行程序。在提示符后输入"ping –a 127.0.0.1"，结果如下：

```
C:\Users\Administrator>ping -a 127.0.0.1
正在 ping DESKTOP-TJCCN0H [127.0.0.1] 具有 32 字节的数据:
来自 127.0.0.1 的回复: 字节=32 时间<1ms TTL=64
来自 127.0.0.1 的回复: 字节=32 时间<1ms TTL=64
来自 127.0.0.1 的回复: 字节=32 时间<1ms TTL=64
来自 127.0.0.1 的回复: 字节=32 时间<1ms TTL=64
127.0.0.1 的 ping 统计信息:
    数据包: 已发送 = 4, 已接收 = 4, 丢失 = 0 (0% 丢失),
往返行程的估计时间(以毫秒为单位):
    最短 = 0ms, 最长 = 0ms, 平均 = 0ms
```

结果显示，在执行 ping 命令的过程中，将目标主机的 IP 地址解析为主机名。

6. ping -i

该参数设置 ICMP 请求报文的 TTL 值为 8，这个值在每经过一台路由器时会被减 1。当被减小到 1 时，路由器会将该分组丢弃，造成超时。所以，当 TTL 值太小时，可能会出现本来网络是通的，但由于 TTL 值耗尽而发生超时的现象，故对此要合理判断。

Windows 系统用户通过执行"win+R"组合键，在弹出的窗口中输入"cmd"来打开命令行程序。在提示符后输入"ping –i 8 www.baidu.com"，结果如下：

```
C:\Users\Administrator>ping -i 8 www.baidu.com
正在 ping www.a.shifen.com [112.80.248.76] 具有 32 字节的数据:
来自 153.3.228.198 的回复: TTL 传输中过期。
来自 153.3.228.198 的回复: TTL 传输中过期。
来自 153.3.228.198 的回复: TTL 传输中过期。
来自 153.3.228.198 的回复: TTL 传输中过期。
112.80.248.76 的 ping 统计信息:
    数据包: 已发送 = 4, 已接收 = 4, 丢失 = 0 (0% 丢失),
```

结果显示，在执行 ping 命令的过程中，可以用域名代替目标主机的 IP 地址，TTL 值设置为 8，在数据包到达 www.baidu.com 服务器之前，该值已减少至 1。

7. 组合参数

Windows 系统用户通过执行"win+R"组合键，在弹出的窗口中输入"cmd"来打开命令行程序。在提示符后输入"ping –a –l 128 –n 2 127.0.0.1"，结果如下：

```
C:\Users\Administrator>ping -a  -l 128 -n 2 127.0.0.1
正在 ping DESKTOP-TJCCN0H [127.0.0.1] 具有 128 字节的数据：
来自 127.0.0.1 的回复：字节=128 时间<1ms TTL=64
来自 127.0.0.1 的回复：字节=128 时间<1ms TTL=64
127.0.0.1 的 ping 统计信息：
    数据包：已发送 = 2，已接收 = 2，丢失 = 0 (0% 丢失)，
往返行程的估计时间(以毫秒为单位)：
    最短 = 0ms，最长 = 0ms，平均 = 0ms
```

结果显示，在执行 ping 命令的过程中，将目标主机的 IP 地址解析为主机名，发送和返回数据包为 2 个 128 字节。

8. 检测网络故障

1) 检测本机 TCP/IP 协议是否工作正常

Windows 系统用户通过执行"win+R"组合键，在弹出的窗口中输入"cmd"来打开命令行程序。在提示符后输入"ping 127.0.0.1"，结果如下：

```
C:\Users\Administrator>ping 127.0.0.1      //本机环回地址
正在 ping 127.0.0.1 具有 32 字节的数据：
来自 127.0.0.1 的回复：字节=32 时间<1ms TTL=64
来自 127.0.0.1 的回复：字节=32 时间<1ms TTL=64
来自 127.0.0.1 的回复：字节=32 时间<1ms TTL=64
来自 127.0.0.1 的回复：字节=32 时间<1ms TTL=64
127.0.0.1 的 ping 统计信息：
    数据包：已发送 = 4，已接收 = 4，丢失 = 0 (0% 丢失)，
往返行程的估计时间(以毫秒为单位)：
    最短 = 0ms，最长 = 0ms，平均 = 0ms
```

结果显示，本机 TCP/IP 协议正常工作。

2) 检测本机网卡是否工作正常

Windows 系统用户通过执行"win+R"组合键，在弹出的窗口中输入"cmd"来打开命令行程序。在提示符后输入"ping 192.168.110.8"，结果如下：

```
C:\Users\Administrator>ping 192.168.110.8      //本机IP地址
```

```
正在 ping 192.168.110.8 具有 32 字节的数据：
来自 192.168.110.8 的回复: 字节=32 时间<1ms TTL=64
来自 192.168.110.8 的回复: 字节=32 时间<1ms TTL=64
来自 192.168.110.8 的回复: 字节=32 时间<1ms TTL=64
来自 192.168.110.8 的回复: 字节=32 时间<1ms TTL=64
192.168.110.8 的 ping 统计信息：
    数据包: 已发送 = 4，已接收 = 4，丢失 = 0 (0% 丢失)，
往返行程的估计时间(以毫秒为单位):
    最短 = 0ms，最长 = 0ms，平均 = 0ms
```

结果显示，本机网卡正常工作。

3) 检测局域网的连通性

Windows 系统用户通过执行"win+R"组合键，在弹出的窗口中输入"cmd"来打开命令行程序。在提示符后输入"ping 192.168.110.1"，结果如下：

```
C:\Users\Administrator>ping 192.168.110.1      //网关IP地址
正在 ping 192.168.110.1 具有 32 字节的数据：
来自 192.168.110.1 的回复: 字节=32 时间=1ms TTL=64
来自 192.168.110.1 的回复: 字节=32 时间=36ms TTL=64
来自 192.168.110.1 的回复: 字节=32 时间=2ms TTL=64
来自 192.168.110.1 的回复: 字节=32 时间=4ms TTL=64
192.168.110.1 的 ping 统计信息：
    数据包: 已发送 = 4，已接收 = 4，丢失 = 0 (0% 丢失)，
往返行程的估计时间(以毫秒为单位):
    最短 = 1ms，最长 = 36ms，平均 = 10ms
```

结果显示，局域网正常工作。

4) 检测 DNS 服务器是否正常工作

Windows 系统用户通过执行"win+R"组合键，在弹出的窗口中输入"cmd"来打开命令行程序。在提示符后输入"ping 202.103.24.68"，结果如下：

```
C:\Users\Administrator>ping 202.103.24.68      //DNS服务器IP地址
正在 ping 192.168.3.1 具有 32 字节的数据：
来自 202.103.24.68 的回复: 字节=32 时间=3ms TTL=64
来自 202.103.24.68 的回复: 字节=32 时间=6ms TTL=64
来自 202.103.24.68 的回复: 字节=32 时间=5ms TTL=64
来自 202.103.24.68 的回复: 字节=32 时间=7ms TTL=64
202.103.24.68 的 ping 统计信息：
    数据包: 已发送 = 4，已接收 = 4，丢失 = 0 (0% 丢失)，
往返行程的估计时间(以毫秒为单位):
```

```
    最短 = 1ms, 最长 = 36ms, 平均 = 10ms
```

还可以在提示符后输入"ping www.baidu.com"来检测 DNS 服务器是否正常工作,因为域名先由域名服务器解析为对应 IP 地址,再执行 ping 命令。

```
C:\Users\Administrator>ping www.baidu.com
正在 ping www.a.shifen.com [112.80.248.75] 具有 32 字节的数据:
来自 112.80.248.75 的回复: 字节=32 时间=24ms TTL=55
来自 112.80.248.75 的回复: 字节=32 时间=23ms TTL=55
来自 112.80.248.75 的回复: 字节=32 时间=21ms TTL=55
来自 112.80.248.75 的回复: 字节=32 时间=21ms TTL=55
112.80.248.75 的 ping 统计信息:
    数据包: 已发送 = 4, 已接收 = 4, 丢失 = 0 (0% 丢失),
往返行程的估计时间(以毫秒为单位):
    最短 = 21ms, 最长 = 24ms, 平均 = 22ms
```

结果显示,DNS 服务器正常工作。

2.3　arp 命令

一、原理概述

arp 命令用来显示和修改 IP 地址与物理地址之间的映射关系,该映射关系用来表示 IP 地址到物理地址之间的转发表,该转发表保存在本地 ARP 缓存中。

Windows 系统用户通过执行"win+R"组合键,在弹出的窗口中输入"cmd"来打开命令行程序。在提示符后,按如下格式输入:

```
ARP -s inet_addr eth_addr [if_addr]
ARP -d inet_addr [if_addr]
ARP -a [inet_addr] [-N if_addr] [-v]
```

arp 命令常见参数及其含义如下:

```
-a            通过询问当前协议数据,显示当前 ARP 项
```

```
-g              与 -a 相同
-v              在详细模式下显示当前 ARP 项。所有无效项和环回接口上的项都将显示
inet_addr       指定 Internet 地址
-N if_addr      显示 if_addr 指定的网络接口的 ARP 项
-d              删除 inet_addr 指定的主机。inet_addr 可以是通配符 *，以删除所有主机
-s              将 Internet 地址 inet_addr与物理地址 eth_addr 相关联。该项是永久的
eth_addr        指定物理地址
```

二、实验目的

1. 理解 arp 命令的基本功能
2. 掌握 ping 命令的几个常用参数使用方法

三、实验内容

在接入局域网的计算机上使用 arp 命令查看和修改 IP 地址与物理地址之间的映射关系。

四、实验步骤

1. arp -a

Windows 系统用户通过执行"win+R"组合键，在弹出的窗口中输入"cmd"来打开命令行程序。在提示符后输入"arp –a"，结果如下：

```
C:\Users\Administrator>arp -a
接口: 192.168.3.39 --- 0xd
  Internet 地址              物理地址                类型
  192.168.3.1                f0-c4-2f-12-1b-dc       动态
  192.168.3.255              ff-ff-ff-ff-ff-ff       静态
  ……
```

结果显示所有已经访问过的局域网主机 IP 地址及对应的物理地址。若只想显示某个指定 IP 地址 192.168.3.1 的 ARP 缓存记录，则可用如下命令：

```
C:\Users\Administrator>arp -a 192.168.3.1
接口: 192.168.3.39 --- 0xd
  Internet 地址              物理地址                类型
  192.168.3.1                f0-c4-2f-12-1b-dc       动态
```

结果只显示了 192.168.3.1 对应的物理地址。

如果计算机有多个接口，若只想显示某个接口的 ARP 缓存记录，可使用"arp –a –n"命令显示某接口的 ARP 缓存记录。

```
C:\Users\Administrator>arp -a -n 192.168.3.39
接口: 192.168.3.39 --- 0xd
  Internet 地址          物理地址              类型
  192.168.3.1           f0-c4-2f-12-1b-dc    动态
  192.168.3.255         ff-ff-ff-ff-ff-ff    静态
  ……
```

2. arp -d

Windows 系统用户通过执行"win+R"组合键，在弹出的窗口中输入"cmd"来打开命令行程序。在提示符后输入"arp –d"，结果如下：

```
C:\Users\Administrator>arp -d
C:\Users\Administrator>arp -a
接口: 192.168.3.39 --- 0xd
  Internet 地址          物理地址              类型
  224.0.0.22            01-00-5e-00-00-16    静态
```

使用 ping 命令测试与网关之间的连通性，再次使用 arp –a 命令显示 ARP 缓存的内容。

```
C:\Users\Administrator>arp -a
接口: 192.168.3.39 --- 0xd
  Internet 地址          物理地址              类型
  192.168.3.1           f0-c4-2f-12-1b-dc    动态
  224.0.0.22            01-00-5e-00-00-16    静态
```

结果显示，使用 arp –d 命令可以删除本地网络 IP 地址对应的 MAC 地址映射项，通过在此访问本地主机（网关），使用 arp –a 命令可见刚刚访问主机 IP 地址对应的 MAC 地址映射项。若只想删除某一个映射项，则可用如下命令：

```
C:\Users\Administrator>arp -d 192.168.3.1
C:\Users\Administrator>arp -a
接口: 192.168.3.39 --- 0xd
  Internet 地址          物理地址              类型
  224.0.0.22            01-00-5e-00-00-16    静态
```

3. arp -s

为了防止 ARP 攻击，即攻击者篡改网关对应的 MAC 地址，可以使用"**arp –s**"命令来添加一条永久的静态映射项。

Windows 系统用户通过执行"**win+R**"组合键，在弹出的窗口中输入"**cmd**"来打开命令行程序。在提示符后输入"**arp –s 192.168.3.1 f0–c4–2f–12–1b–dc**"，结果如下：

```
C:\Users\Administrator>arp -s 192.168.3.1 f0-c4-2f-12-1b-dc
C:\Users\Administrator>arp -a 192.168.3.1
接口: 192.168.3.39 --- 0xd
  Internet 地址           物理地址              类型
  192.168.3.1           f0-c4-2f-12-1b-dc     静态      //此处类型变成了静态
```

使用 ping 命令测试本机与网关之间的连通性，在此使用 arp –a 显示 ARP 缓存的内容。

在一些 Windows 系统中，当运行 arp 命令来添加或删除静态映射项时，有时会被提示"请求的操作需要升级"，这时需要使用管理员身份运行命令行程序。在"开始"处搜索到命令行程序后，右击并选择"以管理员身份运行"，如图 2.1 所示。

图 2.1 以管理员身份运行命令行程序

2.4 netstat 命令

一、原理概述

netstat 命令是 Windows 系统提供的显示协议统计信息和当前 TCP/IP 网络连接的网络工具，并能检验本机各接口的网络连接情况。

Windows 系统用户通过执行"win+R"组合键，在弹出的窗口中输入"cmd"来打开命令行程序。在提示符后，按如下格式输入：

```
netstat [-a][-b][-e][-f][-n][-o][-p proto][-r][-s][-t][-x][-y][interval]
```

netstat 命令常见参数及其含义如下：

```
-a          显示所有连接和侦听端口
-e          显示以太网统计信息。此选项可以与 -s 选项结合使用
-n          以数字形式显示地址和端口号
-o          显示拥有的与每个连接关联的进程 ID
-p proto    显示 proto 指定的协议的连接；proto可是以下任一个：TCP、UDP、TCPv6、UDPv6
```

如果与 -s 选项一起用来显示每个协议的统计信息，proto 可以是下列任何一个：IP、IPv6、ICMP、ICMPv6、TCP、TCPv6、UDP、UDPv6。

```
-r          显示路由表
-s          显示每个协议的统计信息。默认情况下，显示 IP、IPv6、ICMP、ICMPv6、TCP、TCPv6、
            UDP 和 UDPv6 的统计信息
-p          选项可用于指定默认的子网
```

二、实验目的

1. 理解 netstat 命令的基本功能
2. 掌握 netstat 命令的几个常用参数使用方法

三、实验内容

在接入局域网或 Internet 的计算机上使用 netstat 命令显示协议统计信息和当前 TCP/IP 网络连接。

四、实验步骤

1. netstat -a

Windows 系统用户通过执行"win+R"组合键，在弹出的窗口中输入"cmd"来打开命令行程序。在提示符后输入"netstat –a"，结果如下：

第 2 章　常用网络命令

```
C:\Users\Administrator>netstat -a
活动连接
    协议       本地地址              外部地址                   状态
    ……
    TCP   192.168.3.39:139       DESKTOP-TJCCN0H:0       LISTENING
    TCP   192.168.3.39:49976     20.198.162.76:https     ESTABLISHED
    TCP   192.168.3.39:49996     116.181.3.49:http       CLOSE_WAIT
    TCP   192.168.3.39:50021     120.52.190.34:https     CLOSE_WAIT
    TCP   192.168.3.39:50026     139.170.154.232:http    TIME_WAIT
    TCP   192.168.3.39:50029     112.65.193.170:https    TIME_WAIT
    TCP   192.168.3.39:50058     112.65.193.150:https    TIME_WAIT
    TCP   192.168.3.39:50062     112.86.240.82:https     ESTABLISHED
    TCP   192.168.3.39:50063     112.65.193.150:https    TIME_WAIT
    TCP   192.168.3.39:50064     112.80.248.75:http      FIN_WAIT_1
    TCP   192.168.3.39:50065     112.80.248.75:http      FIN_WAIT_1
    TCP   192.168.56.1:139       DESKTOP-TJCCN0H:0       LISTENING
    UDP   [fe80::f819:cb26:9a35:793c%13]:1900     *:*
    UDP   [fe80::f819:cb26:9a35:793c%13]:62385    *:*
    ……
```

结果显示，协议栏显示了所使用的协议，本地地址包括 IP 地址及对应的端口号。外部地址指远程主机地址及端口号，端口号也用协议代替。状态栏显示各连接协议的状态。如果系统正在访问外部资源，该命令还会显示一些活动连接项。

2．netstat -n

Windows 系统用户通过执行 "**win+R**" 组合键，在弹出的窗口中输入 "**cmd**" 来打开命令行程序。在提示符后输入 "**netstat –n**"，结果如下：

```
C:\Users\Administrator>netstat -n
活动连接
    协议   本地地址                  外部地址                   状态
    TCP   127.0.0.1:49736          127.0.0.1:49737         ESTABLISHED
    TCP   127.0.0.1:49737          127.0.0.1:49736         ESTABLISHED
    TCP   192.168.3.39:49976       20.198.162.76:443       ESTABLISHED
    TCP   192.168.3.39:49996       116.181.3.49:80         CLOSE_WAIT
    TCP   192.168.3.39:50020       119.3.178.178:21114     ESTABLISHED
    TCP   192.168.3.39:50021       120.52.190.34:443       CLOSE_WAIT
    TCP   192.168.3.39:50606       122.189.80.3:443        CLOSE_WAIT
    TCP   192.168.3.39:50630       112.86.230.221:443      CLOSE_WAIT
    TCP   192.168.3.39:50641       112.65.193.150:443      TIME_WAIT
    TCP   192.168.3.39:50646       112.65.193.150:443      TIME_WAIT
    TCP   192.168.3.39:50650       112.65.193.150:443      TIME_WAIT
    TCP   192.168.3.39:50659       112.80.248.76:80        FIN_WAIT_1
    TCP   192.168.3.39:50660       116.128.163.191:443     ESTABLISHED
    TCP   192.168.3.39:50661       112.80.248.76:80        FIN_WAIT_1
```

```
TCP    192.168.3.39:50662         112.65.193.150:443         ESTABLISHED
```

结果显示，以数字形式显示地址和端口号，而 netstat –a 显示成 IP 地址和协议。在测试命令前，可以先访问一些 Web 资源，再运行本命令，结果显示活动连接项。

3. netstat -e

Windows 系统用户通过执行"win+R"组合键，在弹出的窗口中输入"cmd"来打开命令行程序。在提示符后输入"netstat –e"，结果如下：

```
C:\Users\Administrator>netstat -e
接口统计
                    接收的          发送的
字节              47331531         10259811
单播数据包          68652           70875
非单播数据包        66321            666
……
```

结果显示关于以太网的统计数据，列出的项目包括传送的数据包的总字节数、错误数、删除数、数据包的单播和广播的数量。这些统计数据既有接收数据包数量，也有发送数据包的数量。

4. netstat -s

Windows 系统用户通过执行"win+R"组合键，在弹出的窗口中输入"cmd"来打开命令行程序。在提示符后输入"netstat –s"，结果如下：

```
C:\Users\Administrator>netstat -s
IPv4 统计信息
接收的数据包               = 114426
  接收的地址错误           = 134
  丢弃的接收数据包         = 3275
  传送的接收数据包         = 144038
  输出请求                 = 116075
  丢弃的输出数据包         = 244
  输出数据包无路由         = 65
……
IPv4 的 TCP 统计信息
  主动开放                 = 5495
  被动开放                 = 27
  失败的连接尝试           = 643
  重置连接                 = 301
  当前连接                 = 13
```

第 2 章 常用网络命令

```
    重新传输的分段              = 11797
IPv4 的 UDP 统计信息
    接收的数据报              = 46689
    无端口                  = 3285
    接收错误                 = 17
    发送的数据报              = 10796
......
```

以上是部分结果，该命令按照各个协议分别显示统计数据，在默认情况下，显示 IPv4 和 IPv6 的统计信息，以及之下的 ICMP、TCP 和 UDP 的统计信息。当访问网络资源出现故障时，可通过查看统计数据，确定问题所在。

5. netstat -r

Windows 系统用户通过执行"win+R"组合键，在弹出的窗口中输入"cmd"来打开命令行程序。在提示符后输入"netstat –r"，结果如下：

```
C:\Users\Administrator>netstat -r
===========================================================================
接口列表
21...ac 87 a3 33 e5 f5 ......Broadcom NetXtreme Gigabit Ethernet
 4...0a 00 27 00 00 04 ......VirtualBox Host-Only Ethernet Adapter
13...34 36 3b cd 4e 8a ......Broadcom 802.11ac Network Adapter
===========================================================================
IPv4 路由表
===========================================================================
活动路由:
网络目标            网络掩码          网关            接口           跃点数
  0.0.0.0          0.0.0.0       192.168.3.1    192.168.3.39      55
192.168.3.0    255.255.255.0      在链路上        192.168.3.39     311
192.168.3.39   255.255.255.255    在链路上        192.168.3.39     311
192.168.3.255  255.255.255.255    在链路上        192.168.3.39     311
224.0.0.0        240.0.0.0       在链路上         127.0.0.1        331
224.0.0.0        240.0.0.0       在链路上        192.168.3.39     311
255.255.255.255 255.255.255.255   在链路上        192.168.3.39     311
===========================================================================
永久路由:    无
......
```

以上是部分结果，该命令显示关于路由表的信息，除了显示有效路由外，还显示当前有效连接。

6. netstat -p tcp

Windows 系统用户通过执行"win+R"组合键，在弹出的窗口中输入"cmd"来打开命令行

程序。在提示符后输入"netstat –p tcp",结果如下:

```
C:\Users\Administrator>netstat -p tcp
活动连接
  协议     本地地址                 外部地址                    状态
  TCP     127.0.0.1:49736         DESKTOP-TJCCN0H:49737      ESTABLISHED
  TCP     127.0.0.1:49737         DESKTOP-TJCCN0H:49736      ESTABLISHED
  TCP     192.168.3.39:63059      20.198.162.78:https        ESTABLISHED
  TCP     192.168.3.39:63084      116.181.3.49:http          CLOSE_WAIT
  TCP     192.168.3.39:63101      ecs-119-3-178-178:21113    ESTABLISHED
  TCP     192.168.3.39:63103      120.52.190.34:https        CLOSE_WAIT
  TCP     192.168.3.39:63138      27.221.123.252:http        TIME_WAIT
  TCP     192.168.3.39:63143      112.65.193.150:https       TIME_WAIT
……
```

该命令显示当前 TCP 连接,并对 TCP 协议进行统计。–p 参数可以是 IPv4 和 IPv6,以及之下的 ICMP、TCP 和 UDP 等。

2.5 tracert 命令

一、原理概述

tracert 命令用于跟踪源站到目的站之间经过的路由。数据报的 TTL 值每经过一台路由器,其值减 1,当 TTL=0 时,向源站报告 TTL 超时。

Windows 系统用户通过执行"**win+R**"组合键,在弹出的窗口中输入"**cmd**"来打开命令行程序。在提示符后,按如下格式输入:

```
tracert [-d][-h maximum_hops][-j host-list][-w timeout][-R][-S srcaddr][-4][-6] target_name
```

tracert 命令常见参数及其含义如下:

```
-d                          不将地址解析成主机名
-h maximum_hops             搜索目标的最大跃点数
-j host-list                与主机列表一起的松散源路由 (仅适用于IPv4)
-w timeout                  等待每个回复的超时时间 (以毫秒为单位)
-R                          跟踪往返行程路径 (仅适用于IPv6)
```

```
-S srcaddr                              要使用的源地址(仅适用于IPv6)
-4                                      强制使用IPv4
```

二、实验目的

1. 理解 tracert 命令的基本功能
2. 掌握 tracert 命令的几个常用参数使用方法

三、实验内容

在接入局域网或 Internet 的计算机上使用 tracert 命令显示协议统计信息和当前 TCP/IP 网络连接。

四、实验步骤

1. tracert www.baidu.com

Windows 系统用户通过执行"win+R"组合键,在弹出的窗口中输入"**cmd**"来打开命令行程序。在提示符后输入"**tracert www.baidu.com**",结果如下:

```
C:\Users\Administrator>tracert www.baidu.com
通过最多 30 个跃点跟踪
到 www.a.shifen.com [112.80.248.76] 的路由:
 1    7 ms    14 ms    15 ms  192.168.3.1
 2   13 ms    12 ms    12 ms  192.168.1.1
 3   14 ms     6 ms     8 ms  100.112.0.1
 4    9 ms    20 ms     8 ms  183.92.129.77
 5   11 ms    16 ms    20 ms  58.19.215.126
 6    *        *        *     请求超时。
 7   30 ms    34 ms    28 ms  219.158.17.74
 8   53 ms    36 ms    33 ms  153.3.228.198
 9   42 ms    31 ms    72 ms  153.37.96.242
10  119 ms    26 ms    22 ms  182.61.216.0
11    *        *        *     请求超时。
12   32 ms    27 ms    23 ms  112.80.248.76
跟踪完成。
```

tracert 命令后面可以跟域名或 IP 地址,默认 TTL 值为 30。结果显示,从本机到 www.baidu.com 之间,经过了 12 个路由,其中有 2 个路由请求超时,原因可能是因为路由器设置了禁止执行该命令等。

2. tracert -h 5 www.baidu.com

Windows 系统用户通过执行"win+R"组合键,在弹出的窗口中输入"cmd"来打开命令行程序。在提示符后输入"tracert -h 5 www.baidu.com",结果如下:

```
C:\Users\Administrator>tracert -h 5 www.baidu.com
通过最多 5 个跃点跟踪
到 www.a.shifen.com [112.80.248.76] 的路由:
  1     2 ms     1 ms     4 ms  192.168.3.1
  2    14 ms     4 ms    20 ms  192.168.1.1
  3    13 ms     9 ms     8 ms  100.112.0.1
  4    12 ms    10 ms     6 ms  183.92.129.77
  5    11 ms    13 ms    16 ms  58.19.215.126
跟踪完成。
```

结果只跟踪到第 5 个路由就跟踪完毕。

第 3 章 组建以太网

3.1 双绞线制作

一、原理概述

双绞线是结构化布线中常见的传输介质,由两条相互绝缘的铜线按照一定的规则相互缠绕而成的,根据外部是否有屏蔽层分为屏蔽双绞线(STP)和非屏蔽双绞线(UTP)。两条铜线之所以缠绕在一起是因为这样可以减少信号之间的串扰。如果外界电磁信号在两条铜线上产生的干扰大小相等而相位相反,那么这个干扰信号就会相互抵消。另外,每对线使用不同的颜色进行区分。

双绞线制作就是将双绞线和 RJ-45(水晶头)连接在一起。双绞线的制作有两种标准,分别是 EIA/TIA-568A 和 EIA/TIA-568B 标准。当双绞线两端同时是 EIA/TIA-568A 或 EIA/TIA-568B 时,称为直连双绞线,用来连接不同设备接口。若两端不一样,则称为交叉双绞线,用来连接相同设备接口。现在一些网卡、交换机或路由器接口可以自适应直连和交叉方式进行通信。因此,直连双绞线应用更加广泛。

EIA/TIA-568A 标准:白绿、绿、白橙、蓝、白蓝、橙、白棕、棕。

EIA/TIA-568B 标准:白橙、橙、白绿、蓝、白蓝、绿、白棕、棕。

二、实验目的

1. 认识水晶头、双绞线、压线钳、测线仪
2. 掌握制作直连双绞线和交叉双绞线的方法

三、实验内容

制作一根长 50 cm 直连双绞线和一根 50 cm 交叉双绞线。

四、实验步骤

1. 实验准备

制作双绞线前,准备好 50~100 cm 的超五类非屏蔽双绞线、RJ-45 水晶头 3~5 个、压线钳 1 把和测线仪 1 个。

2. 剪线

根据实验对双绞线长度要求,使用压线钳的剪切口剪断 50~100 cm 的网线。

3. 剥线

使用压线钳的剥线口将准备好的双绞线的外保护套管划开,注意不要将里面的铜线的绝缘层划破,刀口距离双绞线的端头至少 2 cm。

4. 理线

按照 EIA/TIA-568B 标准将 8 根导线按规定的序号排好,位置如图 3.1 所示。将 8 根导线平坦整齐地平行排列,导线间不要留空隙。用压线钳的剪切口将 8 根导线前端剪断,注意要整齐,剥线的导线长度为 1.5 cm 左右。

图 3.1　EIA/TIA-568B 标准铜线位置

5. 压线

一只手捏住水晶头,将有弹片的一侧向下,有针脚的一端指向远离自己的方向。另一只手捏平双绞线,最左端是第 1 号引脚,最右边是第 8 号引脚,将剪断的双绞线放入 RJ-45 水晶头中,注意要插到底,并使双绞线的外保护层最后应能够在 RJ-45 水晶头内的凹陷处被压实。将水晶头放入压线钳的 8P 压线槽内,用力压紧,这样,RJ-45 水晶头上的 8 根引脚会切破 8 根铜线绝缘层,和里面的铜线压接在一起。压线钳如图 3.2 所示。

第 3 章 组建以太网

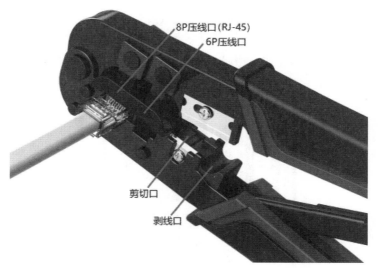

图 3.2 压线钳

用同样的方法制作双绞线的另一端，这样一根直连线就制作好了。

6. 测试双绞线的连通性

测试时将双绞线两端的 RJ-45 水晶头分别插入测线仪的主机和副机的 RJ-45 接口，将开关开至"ON"（S 为慢速显示），若主机和副机的指示灯从 1 至 8 逐一顺序对应闪亮，则直连双绞线制作成功；若有灯不亮，则说明该灯对应的铜线不通。测线仪如图 3.3 所示。

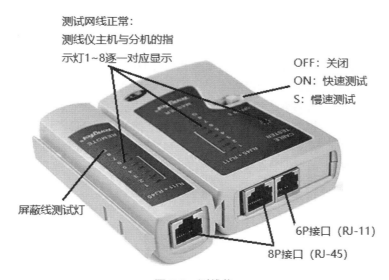

图 3.3 测线仪

7. 制作交叉双绞线

剪掉制作好的直连双绞线的一端，重新按照 EIA/TIA-568A 制作双绞线的方法剪掉一端水

45

晶头，8 根铜线的顺序如图 3.4 所示。制作好交叉双绞线。测试时主机指示灯从 1 至 8 逐一顺序闪亮，若副机指示灯闪亮顺序为：3-6-1-4-5-2-7-8，则说明交叉双绞线制作成功。注意：水晶头是一次性的，制作过程中出现错误，直接剪掉，重新制作新的水晶头。

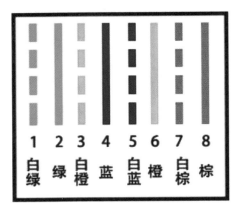

图 3.4　EIA/TIA-568A 标准铜线位置

3.2　集线器和交换机组建以太网

一、原理概述

集线器工作在物理层，每发送一个数据，所有端口均可以收到，采用广播的方式转发数据，因此网络性能受到很大的限制。集线器内部采用了总线型拓扑结构，只能工作在半双工模式下。集线器所连接的局域网称为共享式的以太网，所有端口共享一条带宽。

交换机主要工作在数据链路层，每个端口号及其连接的主机 MAC 地址形成一张转发表，根据数据帧的 MAC 地址转发数据，交换机上的端口之间的通道是相互独立的，可以实现全双工通信。交换机所连接的局域网称为交换式的以太网，所有端口独享带宽。

二、实验目的

1. 理解集线器的工作方式
2. 理解交换机的工作方式
3. 理解交换机地址学习的过程

三、实验内容

本实验模拟一个局域网，其中交换机连接 3 台计算机，集线器连接 3 台计算机，交换机和集线器之间使用直连线相互连接。验证集线器和交换机的工作原理。

四、实验拓扑

集线器和交换机组建以太网拓扑如图 3.5 所示。

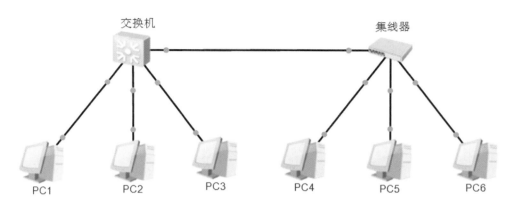

图 3.5　集线器和交换机组建以太网拓扑

五、实验地址分配

本实验地址分配如表 3.1 所示。（MAC 地址随机生成，无须分配）

表 3.1　实验地址分配

设备	接口	MAC 地址	IP 地址	子网掩码	默认网关
PC1	Ethernet 0/0/1	54-89-98-06-44-44	192.168.1.1	255.255.255.0	N/A
PC2	Ethernet 0/0/1	54-89-98-73-56-F4	192.168.1.2	255.255.255.0	N/A
PC3	Ethernet 0/0/1	54-89-98-49-5D-27	192.168.1.3	255.255.255.0	N/A
PC4	Ethernet 0/0/1	54-89-98-17-6D-0F	192.168.1.4	255.255.255.0	N/A
PC5	Ethernet 0/0/1	54-89-98-AD-4C-EB	192.168.1.5	255.255.255.0	N/A
PC6	Ethernet 0/0/1	54-89-98-83-54-1E	192.168.1.6	255.255.255.0	N/A

六、实验步骤

1. PC4~PC6 的基本配置

根据实验地址分配进行相应的基本 IP 地址配置，如图 3.6 所示，以 PC4 设置 IP 地址为 **192.168.1.4**，子网掩码为 **255.255.255.0**，单击"应用"按钮。

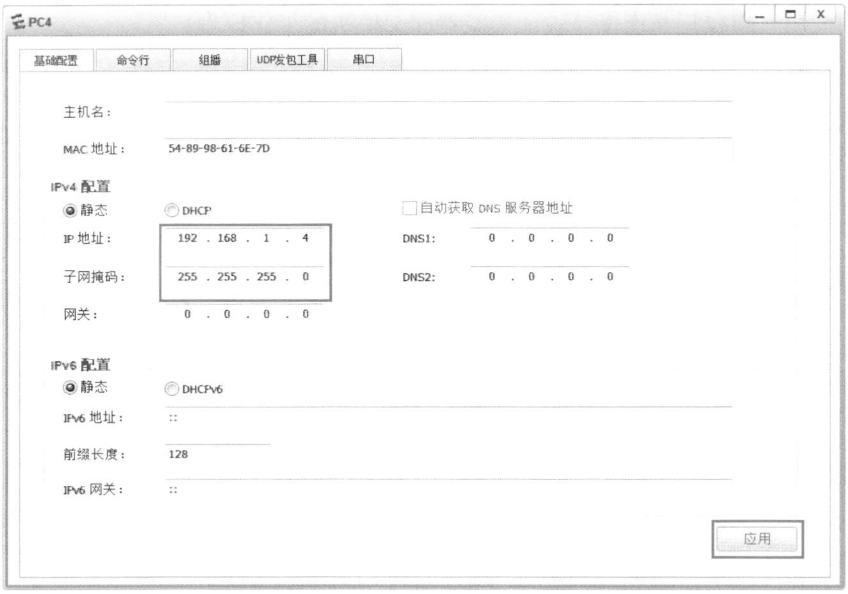

图 3.6　PC4 的基本配置

根据实验地址分配进行相应的基本 IP 地址配置，如图 3.7 所示，以 PC5 设置 IP 地址为 192.168.1.5，子网掩码为 255.255.255.0，单击"应用"按钮。

图 3.7　PC5 的基本配置

根据实验地址分配进行相应的基本 IP 地址配置，如图 3.8 所示，以 PC6 设置 IP 地址为 192.168.1.6，子网掩码为 255.255.255.0，单击"应用"按钮。

图 3.8　PC6 的基本配置

2. 验证集线器的工作原理

在 PC4 上使用 ping 命令测试 PC4 到 PC5 的连通性。同时分别在 PC4 和 PC5 上抓包分析数据流。

```
PC>ping 192.168.1.5
ping 192.168.1.5: 32 data bytes, Press Ctrl_C to break
From 192.168.1.5: bytes=32 seq=1 ttl=128 time=47 ms
From 192.168.1.5: bytes=32 seq=2 ttl=128 time=47 ms
From 192.168.1.5: bytes=32 seq=3 ttl=128 time=47 ms
From 192.168.1.5: bytes=32 seq=4 ttl=128 time=63 ms
From 192.168.1.5: bytes=32 seq=5 ttl=128 time=47 ms
--- 192.168.1.5 ping statistics ---
  5 packet(s) transmitted
  5 packet(s) received
  0.00% packet loss
```

在 PC5 和 PC6 上抓包。在 PC5 上抓包操作如图 3.9 所示。

PC5 上抓包的结果显示，在 PC5 上收到 PC4 上使用 ping 命令（网络层 ICMP）发送和接收的 5 个数据包，如图 3.10 所示。

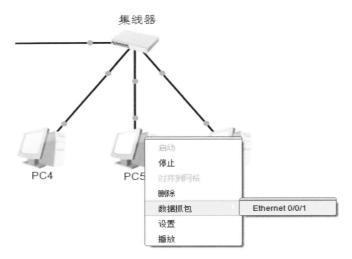

图 3.9　在 PC5 的接口上抓包操作

文件(F)　编辑(E)　视图(V)　跳转(G)　捕获(C)　分析(A)　统计(S)　电话(Y)　无线(W)　工具(T)　帮助(H)

应用显示过滤器 … <Ctrl-/>

No.	Time	Source	Destination	Protocol	Length	Info
58	105.453000	HuaweiTe_17:6d:0f	Broadcast	ARP	60	Who has 192.168.1.5? Tell 192.168.1.4
59	105.469000	HuaweiTe_ad:4c:eb	HuaweiTe_17:6d:0f	ARP	60	192.168.1.5 is at 54:89:98:ad:4c:eb
60	105.484000	192.168.1.4	192.168.1.5	ICMP	74	Echo (ping) request id=0x9f91, seq=1/256, ttl=128 (reply in 61)
61	105.500000	192.168.1.5	192.168.1.4	ICMP	74	Echo (ping) reply id=0x9f91, seq=1/256, ttl=128 (request in 60)
63	106.516000	192.168.1.4	192.168.1.5	ICMP	74	Echo (ping) request id=0xa091, seq=2/512, ttl=128 (reply in 64)
64	106.531000	192.168.1.5	192.168.1.4	ICMP	74	Echo (ping) reply id=0xa091, seq=2/512, ttl=128 (request in 63)
65	107.547000	192.168.1.4	192.168.1.5	ICMP	74	Echo (ping) request id=0xa191, seq=3/768, ttl=128 (reply in 66)
66	107.563000	192.168.1.5	192.168.1.4	ICMP	74	Echo (ping) reply id=0xa191, seq=3/768, ttl=128 (request in 65)
68	108.578000	192.168.1.4	192.168.1.5	ICMP	74	Echo (ping) request id=0xa291, seq=4/1024, ttl=128 (reply in 69)
69	108.594000	192.168.1.5	192.168.1.4	ICMP	74	Echo (ping) reply id=0xa291, seq=4/1024, ttl=128 (request in 68)
70	109.609000	192.168.1.4	192.168.1.5	ICMP	74	Echo (ping) request id=0xa391, seq=5/1280, ttl=128 (reply in 71)
71	109.609000	192.168.1.5	192.168.1.4	ICMP	74	Echo (ping) reply id=0xa391, seq=5/1280, ttl=128 (request in 70)

图 3.10　在 PC5 接口上抓包结果

PC6 上抓包的结果显示，在 PC6 上收到 PC4 上使用 **ping** 命令（网络层 ICMP）发送和接收的 5 个数据包，如图 3.11 所示。

文件(F)　编辑(E)　视图(V)　跳转(G)　捕获(C)　分析(A)　统计(S)　电话(Y)　无线(W)　工具(T)　帮助(H)

应用显示过滤器 … <Ctrl-/>

No.	Time	Source	Destination	Protocol	Length	Info
13	26.781000	HuaweiTe_17:6d:0f	Broadcast	ARP	60	Who has 192.168.1.5? Tell 192.168.1.4
14	26.797000	HuaweiTe_ad:4c:eb	HuaweiTe_17:6d:0f	ARP	60	192.168.1.5 is at 54:89:98:ad:4c:eb
15	26.812000	192.168.1.4	192.168.1.5	ICMP	74	Echo (ping) request id=0x9f91, seq=1/256, ttl=128 (reply in 16)
16	26.828000	192.168.1.5	192.168.1.4	ICMP	74	Echo (ping) reply id=0x9f91, seq=1/256, ttl=128 (request in 15)
18	27.844000	192.168.1.4	192.168.1.5	ICMP	74	Echo (ping) request id=0xa091, seq=2/512, ttl=128 (reply in 19)
19	27.859000	192.168.1.5	192.168.1.4	ICMP	74	Echo (ping) reply id=0xa091, seq=2/512, ttl=128 (request in 18)
20	28.875000	192.168.1.4	192.168.1.5	ICMP	74	Echo (ping) request id=0xa191, seq=3/768, ttl=128 (reply in 21)
21	28.891000	192.168.1.5	192.168.1.4	ICMP	74	Echo (ping) reply id=0xa191, seq=3/768, ttl=128 (request in 20)
23	29.906000	192.168.1.4	192.168.1.5	ICMP	74	Echo (ping) request id=0xa291, seq=4/1024, ttl=128 (reply in 24)
24	29.922000	192.168.1.5	192.168.1.4	ICMP	74	Echo (ping) reply id=0xa291, seq=4/1024, ttl=128 (request in 23)
25	30.937000	192.168.1.4	192.168.1.5	ICMP	74	Echo (ping) request id=0xa391, seq=5/1280, ttl=128 (reply in 26)
26	30.937000	192.168.1.5	192.168.1.4	ICMP	74	Echo (ping) reply id=0xa391, seq=5/1280, ttl=128 (request in 25)

图 3.11　在 PC6 接口上抓包结果

两次抓包结果显示，PC4 与 PC5 之间的数据传输，PC6 也收到该数据。即通过集线器连接的网络，各主机处于同一个冲突域，即 PC4 发送数据，其他主机 PC5、PC6 都收到该数据。

3. PC1~PC3 的基本配置

根据实验地址分配，完成 PC1~PC3 地址配置。

PC1 的配置如图 3.12 所示。

图 3.12 PC1 的基本配置

PC2 的配置如图 3.13 所示。

图 3.13 PC2 的基本配置

PC3 的配置如图 3.14 所示。

图 3.14 PC3 的基本配置

4. 交换机转发表的学习过程

在交换机上使用 display mac-table 命令查看交换机的转发表。

```
<Huawei>system-view
Enter system view, return user view with Ctrl+Z.
[Huawei]display mac-table  //该命令在系统视图下运行
```

```
[Huawei]                         //没有显示结果，说明转发表是空的
```

在 PC1 上使用 ping 命令测试与 PC2 的连通性。同时在 PC2 和 PC3 上抓包。

```
PC>ping 192.168.1.2
ping 192.168.1.2: 32 data bytes, Press Ctrl_C to break
From 192.168.1.2: bytes=32 seq=1 ttl=128 time=47 ms
From 192.168.1.2: bytes=32 seq=2 ttl=128 time=47 ms
From 192.168.1.2: bytes=32 seq=3 ttl=128 time=47 ms
From 192.168.1.2: bytes=32 seq=4 ttl=128 time=31 ms
From 192.168.1.2: bytes=32 seq=5 ttl=128 time=31 ms
--- 192.168.1.2 ping statistics ---
  5 packet(s) transmitted
  5 packet(s) received
  0.00% packet loss
  round-trip min/avg/max = 31/40/47 ms
```

在交换机上使用 display mac-table 命令查看交换机的转发表。

```
[Huawei]display mac-table
MAC address table of slot 0:
-------------------------------------------------------------------------
MAC Address    vlan/    PEvlan CEvlan  Port      Type      LSP/LSR-ID
               VSI/SI                                      MAC-Tunnel
-------------------------------------------------------------------------
5489-9806-4444 1        -      -       GE0/0/1   dynamic   0/-
5489-9873-56f4 1        -      -       GE0/0/2   dynamic   0/-
-------------------------------------------------------------------------
Total matching items on slot 0 displayed = 2
```

同样地，在 PC1 上使用 ping 命令测试与 PC3、PC4、PC5、PC6 之间的连通性。每次测试之后均在交换机上使用 display mac-table 命令查看交换机的转发表，观察转发表的变化，每 ping 一次，转发表里就多出一项。下面是最终的结果。

```
[Huawei]display mac-table
MAC address table of slot 0:
-------------------------------------------------------------------------
MAC Address    vlan/    PEvlan CEvlan  Port      Type      LSP/LSR-ID
               VSI/SI                                      MAC-Tunnel
-------------------------------------------------------------------------
5489-9806-4444 1        -      -       GE0/0/1   dynamic   0/-
5489-9873-56f4 1        -      -       GE0/0/2   dynamic   0/-
5489-9849-5d27 1        -      -       GE0/0/3   dynamic   0/-
```

```
5489-9817-6d0f 1         -     -     GE0/0/4   dynamic   0/-
5489-98ad-4ceb 1         -     -     GE0/0/4   dynamic   0/-
5489-9883-541e 1         -     -     GE0/0/4   dynamic   0/-
-----------------------------------------------------------------------------
Total matching items on slot 0 displayed = 6
```

5. 交换机的工作原理

在 PC1 上使用 ping 命令测试与 PC2 的连通性。同时在 PC2 和 PC3 上抓包。在 PC2 上抓包的结果如图 3.15 所示。

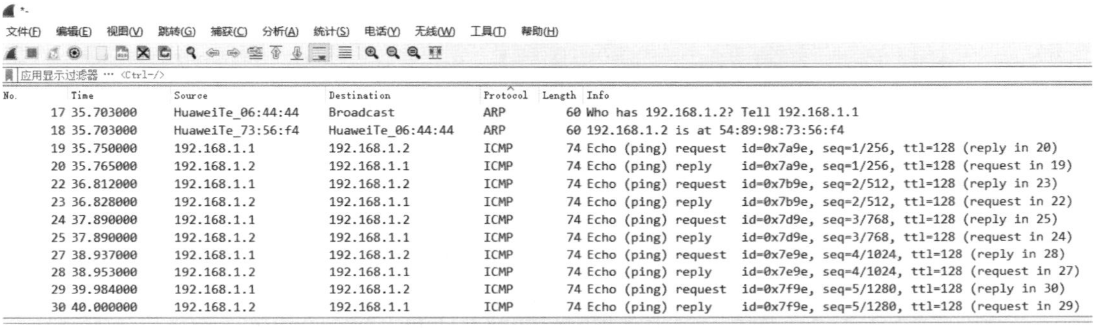

No.	Time	Source	Destination	Protocol	Length	Info
17	35.703000	HuaweiTe_06:44:44	Broadcast	ARP	60	Who has 192.168.1.2? Tell 192.168.1.1
18	35.703000	HuaweiTe_73:56:f4	HuaweiTe_06:44:44	ARP	60	192.168.1.2 is at 54:89:98:73:56:f4
19	35.750000	192.168.1.1	192.168.1.2	ICMP	74	Echo (ping) request id=0x7a9e, seq=1/256, ttl=128 (reply in 20)
20	35.765000	192.168.1.2	192.168.1.1	ICMP	74	Echo (ping) reply id=0x7a9e, seq=1/256, ttl=128 (request in 19)
22	36.812000	192.168.1.1	192.168.1.2	ICMP	74	Echo (ping) request id=0x7b9e, seq=2/512, ttl=128 (reply in 23)
23	36.828000	192.168.1.2	192.168.1.1	ICMP	74	Echo (ping) reply id=0x7b9e, seq=2/512, ttl=128 (request in 22)
24	37.890000	192.168.1.1	192.168.1.2	ICMP	74	Echo (ping) request id=0x7d9e, seq=3/768, ttl=128 (reply in 25)
25	37.890000	192.168.1.2	192.168.1.1	ICMP	74	Echo (ping) reply id=0x7d9e, seq=3/768, ttl=128 (request in 24)
27	38.937000	192.168.1.1	192.168.1.2	ICMP	74	Echo (ping) request id=0x7e9e, seq=4/1024, ttl=128 (reply in 28)
28	38.953000	192.168.1.2	192.168.1.1	ICMP	74	Echo (ping) reply id=0x7e9e, seq=4/1024, ttl=128 (request in 27)
29	39.984000	192.168.1.1	192.168.1.2	ICMP	74	Echo (ping) request id=0x7f9e, seq=5/1280, ttl=128 (reply in 30)
30	40.000000	192.168.1.2	192.168.1.1	ICMP	74	Echo (ping) reply id=0x7f9e, seq=5/1280, ttl=128 (request in 29)

图 3.15　PC2 上抓包结果

结果显示，PC2 收到了 PC1 与 PC2 之间 ICMP 数据。

在 PC3 上抓包的结果如图 3.16 所示。

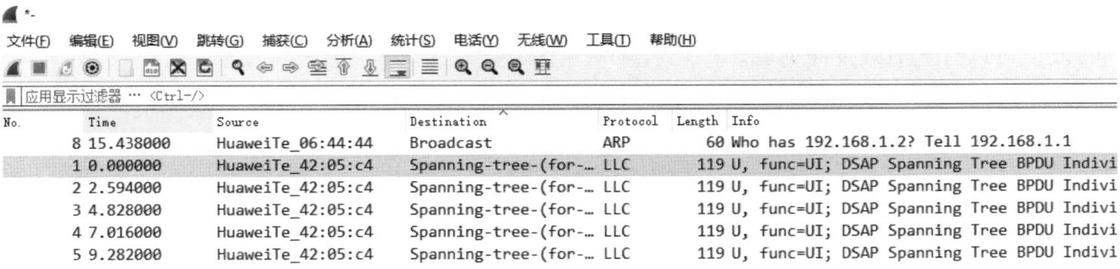

No.	Time	Source	Destination	Protocol	Length	Info
8	15.438000	HuaweiTe_06:44:44	Broadcast	ARP	60	Who has 192.168.1.2? Tell 192.168.1.1
1	0.000000	HuaweiTe_42:05:c4	Spanning-tree-(for-…	LLC	119	U, func=UI; DSAP Spanning Tree BPDU Indivi
2	2.594000	HuaweiTe_42:05:c4	Spanning-tree-(for-…	LLC	119	U, func=UI; DSAP Spanning Tree BPDU Indivi
3	4.828000	HuaweiTe_42:05:c4	Spanning-tree-(for-…	LLC	119	U, func=UI; DSAP Spanning Tree BPDU Indivi
4	7.016000	HuaweiTe_42:05:c4	Spanning-tree-(for-…	LLC	119	U, func=UI; DSAP Spanning Tree BPDU Indivi
5	9.282000	HuaweiTe_42:05:c4	Spanning-tree-(for-…	LLC	119	U, func=UI; DSAP Spanning Tree BPDU Indivi

图 3.16　PC3 上抓包结果

结果显示，PC3 没有收到 PC1 与 PC2 之间的 ICMP 数据。说明，PC1 发送到 PC2 的数据，PC3 并没有收到。

第 4 章 交换机配置

4.1 交换机基本配置

一、原理概述

交换机（switch Hub）拥有一条很高带宽的背部总线和内部交换矩阵。交换机的所有端口都挂接在这条背部总线上，控制电路收到数据包以后，处理端口会查找内存中的地址对照表以确定目的 MAC（网卡的硬件地址）的 NIC（网卡）挂接在哪个端口上，通过内部交换矩阵迅速将数据包传送到目的端口，目的 MAC 若不存在，广播到所有的端口，接收端口回应后交换机会"学习"新的 MAC 地址，并把它添加到内部 MAC 地址表中。使用交换机也可以把网络"分段"，通过对照 IP 地址表，交换机只允许必要的网络流量通过交换机。通过交换机的过滤和转发，可以有效地减少冲突域。

二、实验目的

1. 掌握查看 MAC 地址映射表的方法
2. 理解双工模式和接口速率
3. 掌握更改双工模式的配置
4. 掌握更改接口速率的配置

三、实验内容

某公司新组建网络，购置了 3 台交换机，其中 SW2 和 SW3 为接入层交换机，SW1 为汇聚层交换机。现在需要对 3 台交换机进行配置，保证交换机的接口使用全双工模式，并根据需要配置接口速率。

四、实验拓扑

交换机基本配置拓扑如图 4.1 所示。

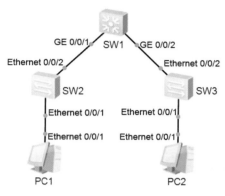

图 4.1　交换机基本配置拓扑

五、实验地址分配

实验地址分配如表 4.1 所示。

表 4.1　实验地址分配

设备	接口	IP 地址	子网掩码	默认网关
PC1	Ethernet 0/0/1	192.168.1.1	255.255.255.0	N/A
PC2	Ethernet 0/0/1	192.168.1.2	255.255.255.0	N/A

六、实验步骤

1. 基本配置

1) PC1 的基本配置

PC1 的基本配置如图 4.2 所示。

图 4.2　PC1 的基本配置

2) PC2 的基本配置

PC2 的基本配置如图 4.3 所示。

IPv4 配置

- 静态 ○DHCP □自动获取 DNS 服务器地址
- IP 地址: 192.168.1.2   DNS1: 0.0.0.0
- 子网掩码: 255.255.255.0   DNS2: 0.0.0.0
- 网关:

图 4.3　PC2 的基本配置

3) 测试

在 PC1 上使用 **ping** 命令检测与 PC2 的连通性。

```
PC>ping 192.168.1.2
ping 192.168.1.2: 32 data bytes, Press Ctrl_C to break
From 192.168.1.2: bytes=32 seq=1 ttl=128 time=93 ms
From 192.168.1.2: bytes=32 seq=2 ttl=128 time=78 ms
From 192.168.1.2: bytes=32 seq=3 ttl=128 time=78 ms
From 192.168.1.2: bytes=32 seq=4 ttl=128 time=78 ms
From 192.168.1.2: bytes=32 seq=5 ttl=128 time=78 ms
--- 192.168.1.2 ping statistics ---
  5 packet(s) transmitted
  5 packet(s) received
  0.00% packet loss
  round-trip min/avg/max = 78/81/93 ms
```

结果显示，PC1 与 PC2 之间通信正常。

2. 设置交换机登录密码

华为交换机密码设置，包括 console 接口密码设置和 telnet 远程登录密码设置，华为所有交换(路由)产品设置方法基本一致。telnet 远程登录密码的设置将在 10.5 节介绍，下面介绍 console 接口密码的设置。

首先通过 consle 接口连接华为交换机，华为产品 consle 默认有初始密码，第一次登录需要修改，在产品说明书查看密码。

根据本实验拓扑中交换机的名称，分别在 3 台交换机的系统视图下使用 **sysname** 命令重新命名交换机为 SW1、SW2 和 SW3。

```
[Huawei]sysname SW1
[SW1]

[Huawei]sysname SW2
```

```
[SW2]
[Huawei]sysname SW3
[SW3]
```

在交换机 SW1 上使用 user-interface console 0 命令切换到 console 0 接口，在 console 0 接口视图中使用 authentication-mode password 命令设置 console 认证方式为密码认证，紧接着使用 set authentication password cipher 123 命令将认证密码设置为 123。

```
[SW1]user-interface console 0
[SW1-ui-console0]authentication-mode password
[SW1-ui-console0]set authentication password cipher 123
```

使用 quit 命令退出交换机重新登录，系统提示输入认证密码，输入认证密码 123，在输入过程中屏幕没有显示输入的痕迹，这是因为我们使用了加密的密码。

```
<SW1>quit                //退出交换机重新登录
<SW1> User interface con0 is available
Please Press ENTER.
Login authentication
Password:              //输入认证密码123
<SW1>
```

结果显示，输入认证密码 123 后，成功登录到用户视图。

3. 查看交换机的 MAC 地址映射表

在 SW2 的用户视图中使用 display mac-address 命令查看 SW1 的 MAC 地址映射表。在 SW1 和 SW3 上可以执行相同的命令查看各自的 MAC 地址映射表。

```
<SW2>display mac-address
MAC address table of slot 0:
-------------------------------------------------------------------
MAC Address      vlan/    PEvlan  CEvlan   Port          Type       LSP/LSR-ID
                 VSI/SI                                  MAC-Tunnel
-------------------------------------------------------------------
5489-9822-2772   1        -       -        Eth0/0/1      dynamic    0/-
5489-9848-3146   1        -       -        Eth0/0/2      dynamic    0/-
-------------------------------------------------------------------
Total matching items on slot 0 displayed = 2
```

4. 配置交换机双工模式

配置接口的双工模式可在自协商或者非自协商模式下进行。自协商模式下，接口的双工模式是对端接口协商得到的，但协商得到的双工模式可能与实际需求不符，可通过配置双工模式的取值范围来控制协商的结果。例如，对端接口都支持半双工/全双工，协商结果为半双工模式，与需要的全双工模式不符，这时就可以执行 **auto duplex full** 命令使接口的协商双工模式变为全双工模式。在非自协商模式下，可根据实际需求手动配置接口的双工模式。

在 SW1、SW2、SW3 交换机接口下先通过 undo negotiation auto 命令关掉自协商功能，再手动指定双工模式为全双工模式。

```
[SW2]interface Ethernet 0/0/2
[SW2-Ethernet0/0/2]undo negotiation auto
[SW2-Ethernet0/0/2]auto duplex full
```

用同样的方法配置另外两台交换机接口的双工模式。

5. 配置接口的速率

在自协商模式下，以太网接口的速率是和对端接口协商得到的，如果协商的速率与实际需求不符，可通过配置接口速率的取值范围来控制协商结果。例如，对端接口都支持 10 Mb/s 和 100 Mb/s 的速率，协商后的速率为 10 Mb/s，与实际需求的 100 Mb/s 不符，可通过执行 speed 100 命令配置使用接口可协商的速率为 100 Mb/s。

在 SW2 接口下配置速率。首先关闭自协商模式，然后配置以太网接口的速率。

```
[SW2]interface Ethernet 0/0/2
[SW2-Ethernet0/0/2]undo negotiation auto
[SW2-Ethernet0/0/2]speed 100
```

用同样的方法配置另外两台交换机接口的速率。

4.2 STP 配置

一、原理概述

STP 是用来避免数据链路层出现逻辑环路的协议，运行 STP 协议的设备通过交互信息发现环路，并通过阻塞特定端口，最终将网络结构修剪成无环路的树型结构。在网络出现故障的时候，STP 能快速发现链路故障，并尽快找出另外一条路径进行数据传输。

第 4 章　交换机配置

交换机上运行的 STP 通过 BPDU 信息的交互，选举根交换机，然后每台非根交换机选择用来与根交换机通信的根端口，之后每个网段选择用来转发数据至根交换机指定的端口，最后剩余的端口则被阻塞。

二、实验目的

1. 理解 STP 的选举过程
2. 掌握修改交换机优先级的方法
3. 掌握修改端口开销值的方法

三、实验内容

公司购置了 4 台交换机，组建网络。由于默认情况下，交换机之间运行 STP 后，根交换机、根端口、指定端口的选择将基于交换机的 MAC 地址的大小，因此带来了不确定性，极可能由此产生隐患。

根据公司网络规划，需要 SW1 作为根交换机，SW2 作为 SW1 的备份根交换机。同时对于 SW4 交换机，Ethernet 0/0/1 接口应该作为根端口。对于 SW2 和 SW3 之间的链路，应该保证 SW2 的 Ethernet 0/0/3 接口作为指定端口。同时在交换机 SW3 上，存在两个接口 Ethernet 0/0/4、Ethernet 0/0/5 连接到测试 PC，测试 PC 经常上下线网络，需要将交换机 SW3 与之相连的对应端口定义为边缘端口，避免测试计算机上下线对网络产生的影响。

四、实验拓扑

STP 配置拓扑如图 4.4 所示。

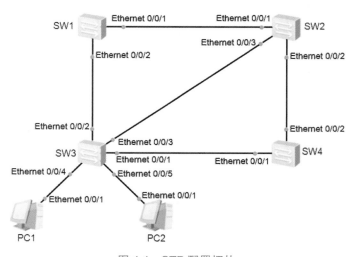

图 4.4　STP 配置拓扑

五、实验地址记录

本实验的 MAC 地址如表 4.2 所示。

表 4.2　MAC 地址表

设备	全局 MAC 地址
SW1（S3700）	4c1f-cc0c-5f8c
SW2（S3700）	4c1f-cc89-232a
SW3（S3700）	4c1f-cc0c-3111
SW4（S3700）	4c1f-cc09-797a

六、实验步骤

1. 基本配置

在各交换机上启动 STP，将交换机的 STP 模式更改为普通生成树 STP。

```
[SW1]stp enable
[SW1]stp mode stp

[SW2]stp enable
[SW2]stp mode stp

[SW3]stp enable
[SW3]stp mode stp

[SW4]stp enable
[SW4]stp mode stp
```

配置完成后，默认情况下需要等待 30 s 生成树重新计算的时间，再使用 **display stp** 命令查看 SW1 的生成树状态。

```
[SW1]display stp
-------[CIST Global Info][Mode STP]-------
CIST Bridge          :32768.4c1f-cc0c-5f8c
……
----[Port1(Ethernet0/0/1)][DISCARDING]----
 Port Protocol         :Enabled
 Port Role             :Alternate Port
 Port Priority         :128
 Port Cost(Dot1T )     :Config=auto / Active=200000
 Designated Bridge/Port   :32768.4c1f-cc89-232a / 128.1
```

```
......
----[Port2(Ethernet0/0/2)][FORWARDING]----
 Port Protocol          :Enabled
 Port Role              :Root Port
 Port Priority          :128
 Port Cost(Dot1T )      :Config=auto / Active=200000
 Designated Bridge/Port :32768.4c1f-cc0c-3111 / 128.2
......
```

结果显示 SW1 的 Ethernet 0/0/1 端口为丢弃状态，端口角色为 Alternate，即替代端口。Ethernet 0/0/2 端口为转发状态，端口角色为根端口。

还可以使用 **display stp brief** 命令在 SW2、SW3、SW4 上仅查看摘要信息。

```
[SW2]display stp brief
 MSTID   Port            Role    STP State    Protection
   0     Ethernet0/0/1   DESI    FORWARDING   NONE
   0     Ethernet0/0/2   ROOT    FORWARDING   NONE
   0     Ethernet0/0/3   ALTE    DISCARDING   NONE
```

在交换机 SW2 上的 Ethernet 0/0/1 和 Ethernet 0/0/2 端口为转发状态，Ethernet 0/0/3 端口为丢弃状态。Ethernet 0/0/1 角色为指定端口，Ethernet 0/0/2 为根端口。

```
[SW3]display stp brief
 MSTID   Port            Role    STP State    Protection
   0     Ethernet0/0/1   ROOT    FORWARDING   NONE
   0     Ethernet0/0/2   DESI    FORWARDING   NONE
   0     Ethernet0/0/3   DESI    FORWARDING   NONE
   0     Ethernet0/0/4   DESI    FORWARDING   NONE
   0     Ethernet0/0/5   DESI    FORWARDING   NONE
```

在交换机 SW3 上，所有端口为转发状态，其中 Ethernet 0/0/1 端口角色为根端口，其他端口为指定端口。

```
[SW4]display stp brief
 MSTID   Port            Role    STP State    Protection
   0     Ethernet0/0/1   DESI    FORWARDING   NONE
   0     Ethernet0/0/2   DESI    FORWARDING   NONE
```

在交换机 SW4 上，所有端口为转发状态，角色为指定端口。

可以初步判断 4 台交换机中 SW4 为根交换机，因为该交换机所有端口都为指定端口。通过 display stp 命令查看生成树详细信息。

```
[SW4]display stp
-------[CIST Global Info][Mode STP]-------
CIST Bridge          :32768.4c1f-cc09-797a
Config Times         :Hello 2s MaxAge 20s FwDly 15s MaxHop 20
Active Times         :Hello 2s MaxAge 20s FwDly 15s MaxHop 20
CIST Root/ERPC       :32768.4c1f-cc09-797a / 0
CIST RegRoot/IRPC    :32768.4c1f-cc09-797a / 0
……
```

结果显示"CIST Root"和"CIST Bridge"相同，即目前根交换机 ID 与自身的交换机 ID 相同，说明目前 SW4 为根交换机。

生成树运算第一步就是通过比较每台交换机的 ID 选举根交换机。交换机 ID 由交换机优先级和 MAC 地址组成，首先比较交换机优先级，数值低的为根交换机；如果优先级相同，则比较 MAC 地址，同样数值低的选举为根交换机。

2. 配置网络中的根交换机

根交换机在网络中的位置非常重要，如果选择了一台性能较差的交换机，或者是部署在接入层的交换机作为根交换机，会影响到整个网络的通信质量及数据传输。所以确定根交换机的位置极为重要。根交换机选举依据是根交换机 ID，值越小越优先，交换机默认的优先级为 32768，该值是可以修改的。

现在将 SW1 配置为主根交换机，SW2 为备份根交换机，将 SW1 的优先级设为 0，SW2 的优先级设为 4096。

```
[SW1]stp priority 0

[SW2]stp priority 4096
```

配置完成后查看 SW1 和 SW2 的 STP 状态信息。

```
[SW1]display stp
-------[CIST Global Info][Mode STP]-------
CIST Bridge          :0    .4c1f-cc0c-5f8c
Config Times         :Hello 2s MaxAge 20s FwDly 15s MaxHop 20
Active Times         :Hello 2s MaxAge 20s FwDly 15s MaxHop 20
CIST Root/ERPC       :0    .4c1f-cc0c-5f8c / 0
CIST RegRoot/IRPC    :0    .4c1f-cc0c-5f8c / 0
……

[SW2]display stp
```

```
-------[CIST Global Info][Mode STP]-------
CIST Bridge              :4096  .4c1f-cc89-232a
Config Times             :Hello 2s MaxAge 20s FwDly 15s MaxHop 20
Active Times             :Hello 2s MaxAge 20s FwDly 15s MaxHop 20
CIST Root/ERPC           :0     .4c1f-cc0c-5f8c / 200000
CIST RegRoot/IRPC        :4096  .4c1f-cc89-232a / 0
……
```

通过观察发现 SW1 的优先级为 0，为根交换机；而 SW2 的优先级变为 4096，为备份根交换机。这里还可以使用另外一种方式配置主根交换机和备份根交换机，首先删除在 SW1 和 SW2 上所配置的优先级，在 SW1 上使用 **stp root primary** 命令配置主根交换机，在 SW2 上使用 **stp root secondary** 命令配置备份根交换机。

```
[SW1]undo stp priority
[SW1]stp root primary

[SW2]undo stp priority
[SW2]stp root secondary
```

配置完成后查看 STP 的状态信息，与前一种方法得到的结果一致，SW1 自动更改优先级为 0，而 SW2 更改为 4096。

3. 根端口的选举结果

生成树在选举出根交换机后，将在每台非根交换机上选举根端口。选举时首先比较交换机上每个端口到达根交换机的路径开销，路径开销最小的端口将成为根端口。如果根路径开销值相同，则比较每个端口所在链路上的上行交换机 ID，如果该交换机 ID 也相同，则比较每个端口所在链路上的上行端口 ID。每台交换机上只能拥有一个根端口。

目前 SW1 为主根交换机，而 SW2 为备份根交换机，查看 SW4 上的生成树信息。

```
[SW4]display stp brief
MSTID   Port                 Role  STP State    Protection
  0     Ethernet0/0/1        ALTE  DISCARDING   NONE
  0     Ethernet0/0/2        ROOT  FORWARDING   NONE
```

结果显示，现在 SW4 的 Ethernet 0/0/2 为根端口，状态为转发状态。SW4 在选举根端口时，首先比较根路径开销，由于拓扑中所有链路都是相同的百兆以太网链路，SW4 经过 SW3 到 SW1 与经过 SW2 到 SW1 的开销值相同；接下来比较 SW4 的两台上行链路的交换机 SW2 和 SW3 的交换机标识，SW2 目前的交换机优先级为 4096，而 SW3 为默认的 32768，所以 SW2 连接的 Ethernet 0/0/2 接口被选为根端口。

统一使用默认路径开销。华为交换机默认的路径开销计算标准使用的是标准的 Dot1T。GE

接口默认路径开销是 20000,而 Ethernet 接口默认路径开销是 200000。查看 SW4 的 Ethernet 0/0/2 接口开销值。

```
<SW4>display stp interface Ethernet 0/0/2
……
----[Port2(Ethernet0/0/2)][FORWARDING]----
 Port Protocol         :Enabled
 Port Role             :Root Port
 Port Priority         :128
 Port Cost(Dot1T )     :Config=auto / Active=200000
……
```

结果显示,接口路径开销采用 Dot1T 的计算方法,Config 是指手工配置的路径开销,Active 是实际使用的路径开销,开销值为 200000。配置 SW4 的 Ethernet 0/0/2 接口的代价值为 200001,即增加该接口的代价值。配置完再次查看该接口的开销值及 STP 状态摘要信息。

```
[SW4]int Ethernet 0/0/2
[SW4-Ethernet0/0/2]stp cost 200001
……
<SW4>display stp interface Ethernet 0/0/2
----[Port2(Ethernet0/0/2)][DISCARDING]----
 Port Protocol         :Enabled
 Port Role             :Alternate Port
 Port Priority         :128
 Port Cost(Dot1T )     :Config=200001 / Active=200001
……
<SW4>display stp brief
 MSTID  Port              Role     STP State    Protection
  0     Ethernet0/0/1     ROOT     LEARNING     NONE
  0     Ethernet0/0/2     ALTE     DISCARDING   NONE
```

发现 Ethernet 0/0/1 端口角色变成根端口,而 Ethernet 0/0/2 变成 Alternate 端口。

4. 指定端口的选择结果

生成树协议在每台非根交换机选举出根端口后,将在每个网段上选举指定端口,选举的比较规则和选举根端口类似。

在选举指定端口时,首先比较两个端口发送与接收 BPDU 中的根路径开销,如果相同,接着比较端口发送与接收 BPDU 交换机 ID,如果优先级相同,需要进一步比较交换机 MAC 地址,最终选出该物理网段的指定端口。

4.3　VLAN 基本配置

一、原理概述

为了避免冲突，同时扩展传统局域网以接入更多计算机，可以在局域网中使用二层交换机。交换机能够隔离冲突，但所有计算机仍处在同一个广播域，降低了网络的效率，而且降低了安全性，即广播域和网络安全问题仍然存在。为了减少广播，提高安全性，可以使用虚拟局域网（VLAN）技术把一个物理的 LAN 逻辑地划分为多个广播域。同一个 VLAN 内的主机之间可以直接通信，而不同 VLAN 的主机之间不能直接通信。广播信息被限制在一个 VLAN 内，同时提高了网络安全性。不同 VLAN 使用不同的 VLAN ID 区分，比较 VLAN ID 是否与自己的 PVID 相同，若相同，则去掉 VLAN 标签后发送该数据帧给主机；若不同，则直接丢弃该数据帧。

二、实验目的

1. 理解 VLAN 的应用场景
2. 掌握 VLAN 的基本配置
3. 掌握 Access 接口的配置方法
4. 掌握 Access 接口加入相应 VLAN 的方法
5. 掌握 Trunk 端口的配置
6. 掌握 Trunk 端口允许特定 VLAN 通过的配置方法
7. 掌握 Trunk 端口允许所有 VLAN 通过的配置方法

三、实验内容

本实验模拟企业网络场景。公司内网是一个大的局域网，二层交换机 SW1 放置在一楼，在一楼办公的部门有 IT 部和人事部；二层交换机 SW2 放置在二楼，在二楼办公的部门有市场部和研发部。由于交换机组成的是一个广播网，交换机连接的所有主机都能互相通信，而公司策略是：不同部门之间主机不能互相通信，同一部门内的主机才可以互相访问。因此，需要在交换机上划分不同的 VLAN，并将连接主机的交换机接口配置成 Access 接口并划分到相应 VLAN 内，配置交换机之间的 Trunk 接口。

四、实验拓扑

VLAN 基本配置拓扑如图 4.5 所示。

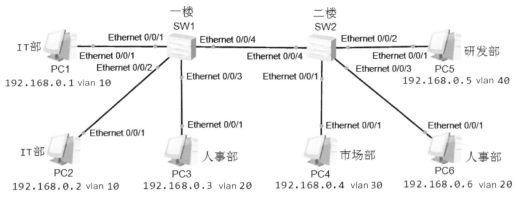

图 4.5　VLAN 基本配置拓扑

五、实验地址分配

本实验的 IP 地址分配如表 4.3 所示。

表 4.3　IP 地址分配表

设备	接口	IP 地址	子网掩码	默认网关	VLAN ID
PC1	Ethernet 0/0/1	192.168.0.1	255.255.255.0	N/A	vlan 10
PC2	Ethernet 0/0/1	192.168.0.2	255.255.255.0	N/A	vlan 10
PC3	Ethernet 0/0/1	192.168.0.3	255.255.255.0	N/A	vlan 20
PC4	Ethernet 0/0/1	192.168.0.4	255.255.255.0	N/A	vlan 30
PC5	Ethernet 0/0/1	192.168.0.5	255.255.255.0	N/A	vlan 40
PC6	Ethernet 0/0/1	192.168.0.6	255.255.255.0	N/A	vlan 20

六、实验步骤

1. 基本配置

根据实验地址分配进行相应的基本 IP 地址配置，在此步骤中不要为交换机创建任何的 VLAN。使用 ping 命令检测各自连通性，所有的 PC 都能相互通信。

```
PC>ping 192.168.0.6
ping 192.168.0.6: 32 data bytes, Press Ctrl_C to break
From 192.168.0.6: bytes=32 seq=1 ttl=128 time=31 ms
From 192.168.0.6: bytes=32 seq=2 ttl=128 time=63 ms
From 192.168.0.6: bytes=32 seq=3 ttl=128 time=47 ms
From 192.168.0.6: bytes=32 seq=4 ttl=128 time=63 ms
From 192.168.0.6: bytes=32 seq=5 ttl=128 time=62 ms
--- 192.168.0.6 ping statistics ---
  5 packet(s) transmitted
```

第 4 章　交换机配置

```
  5 packet(s) received
  0.00% packet loss
  round-trip min/avg/max = 31/53/63 ms
```

2. 创建 VLAN

除默认 vlan 1 外，其余 VLAN 需要通过命令来手工创建。创建 VLAN 有两种方式：一种是使用 vlan 命令一次创建单个 VLAN；另一种方式是使用 vlan batch 命令创建多个 VLAN。

在 SW1 上使用两条命令分别创建 vlan 10 和 vlan 20。

```
[SW1]vlan 10
[SW1-vlan10]vlan 20
```

在 SW2 上使用 vlan batch 命令创建 vlan 20、vlan 30 和 vlan40。

```
[SW2]vlan batch 20 30 40
```

配置完成后，在 SW1 上使用 display vlan 命令查看 VLAN 的相关信息。

```
[SW1]display vlan
The total number of vlans is : 3
--------------------------------------------------------------------------------
U: Up;         D: Down;         TG: Tagged;         UT: Untagged;
MP: vlan-mapping;               ST: vlan-stacking;
#: ProtocolTransparent-vlan;    *: Management-vlan;
--------------------------------------------------------------------------------
VID  Type   Ports
--------------------------------------------------------------------------------
......
10   common
20   common
```

配置完成后，在 SW2 上使用 display vlan 命令查看 VLAN 的相关信息。

```
[SW2]display vlan
The total number of vlans is : 4
--------------------------------------------------------------------------------
U: Up;         D: Down;         TG: Tagged;         UT: Untagged;
MP: vlan-mapping;               ST: vlan-stacking;
#: ProtocolTransparent-vlan;    *: Management-vlan;
--------------------------------------------------------------------------------
```

```
VID  Type    Ports
--------------------------------------------------------------------------
......
20   common
30   common
40   common
```

结果显示，SW1 和 SW2 都已经成功创建了相应 VLAN，但目前没有任何接口加入所创建的 vlan 10 和 vlan 20 中，默认情况下交换机上所有接口都属于 vlan 1。

3. 配置 Access 接口

按照拓扑，使用 port link-type access 命令配置所有 SW1 和 SW2 交换机上连接 PC 的接口类型为 Access 类型接口，并使用 port default vlan 命令配置接口的默认 VLAN 并同时加入相应 VLAN 中。默认情况下，所有接口的默认 VLAN ID 为 1。

```
[SW1]interface Ethernet 0/0/1
[SW1-Ethernet0/0/1]port link-type access
[SW1-Ethernet0/0/1]port default vlan 10
[SW1-Ethernet0/0/1]interface Ethernet 0/0/2
[SW1-Ethernet0/0/2]port link-type access
[SW1-Ethernet0/0/2]port default vlan 10
[SW1-Ethernet0/0/2]interface Ethernet 0/0/3
[SW1-Ethernet0/0/3]port link-type access
[SW1-Ethernet0/0/3]port default vlan 20

[SW2]interface Ethernet 0/0/1
[SW2-Ethernet0/0/1]port link-type access
[SW2-Ethernet0/0/1]port default vlan 30
[SW2-Ethernet0/0/1]interface Ethernet 0/0/2
[SW2-Ethernet0/0/2]port link-type access
[SW2-Ethernet0/0/2]port default vlan 40
[SW2-Ethernet0/0/2]interface Ethernet 0/0/3
[SW2-Ethernet0/0/3]port link-type access
[SW2-Ethernet0/0/3]port default vlan 20
```

配置完成后，查看 SW1 和 SW2 上的 VLAN 信息。

```
[SW1]display vlan
The total number of vlans is : 3
--------------------------------------------------------------------------
U: Up;         D: Down;         TG: Tagged;         UT: Untagged;
MP: vlan-mapping;                ST: vlan-stacking;
#: ProtocolTransparent-vlan;     *: Management-vlan;
--------------------------------------------------------------------------
```

```
VID  Type    Ports
--------------------------------------------------------------
......
10   common  UT:Eth0/0/1(U)      Eth0/0/2(U)
20   common  UT:Eth0/0/3(U)

[SW2]display vlan
The total number of vlans is : 4
--------------------------------------------------------------
U: Up;         D: Down;         TG: Tagged;         UT: Untagged;
MP: vlan-mapping;               ST: vlan-stacking;
#: ProtocolTransparent-vlan;    *: Management-vlan;
--------------------------------------------------------------
VID  Type    Ports
--------------------------------------------------------------
......
20   common  UT:Eth0/0/3(U)
30   common  UT:Eth0/0/1(U)
40   common  UT:Eth0/0/2(U)
```

4. 检查配置结果

在交换机上将不同接口加入各自不同的 VLAN 后，属于相同 VLAN 的接口处于同一个广播域，相互之间可以直接通信。属于不同 VLAN 的接口是处于不同的广播域，相互之间不能直接通信。

在本实验环境中，只有同属于 IT 部门 vlan10 的两台主机 PC1 和 PC2 之间可以互相通信，而与其他部门的 PC 之间无法通信。

在 IT 部门的终端 PC1 上分别测试与同部门的终端 PC2、人事部门的 PC3 之间的连通性。

```
PC>ping 192.168.0.2
ping 192.168.0.2: 32 data bytes, Press Ctrl_C to break
From 192.168.0.2: bytes=32 seq=1 ttl=128 time=47 ms
From 192.168.0.2: bytes=32 seq=2 ttl=128 time=31 ms
From 192.168.0.2: bytes=32 seq=3 ttl=128 time=31 ms
From 192.168.0.2: bytes=32 seq=4 ttl=128 time=31 ms
From 192.168.0.2: bytes=32 seq=5 ttl=128 time=32 ms
--- 192.168.0.2 ping statistics ---
  5 packet(s) transmitted
  5 packet(s) received
  0.00% packet loss
  round-trip min/avg/max = 31/34/47 ms

PC>ping 192.168.0.3
ping 192.168.0.3: 32 data bytes, Press Ctrl_C to break
From 192.168.0.1: Destination host unreachable
From 192.168.0.1: Destination host unreachable
```

```
From 192.168.0.1: Destination host unreachable
From 192.168.0.1: Destination host unreachable
From 192.168.0.1: Destination host unreachable
--- 192.168.0.3 ping statistics ---
 5 packet(s) transmitted
 0 packet(s) received
 100.00% packet loss
```

结果显示，相同 VLAN 内的 PC 之间可以互相通信，不同 VLAN 内的 PC 之间无法通信。但同属于人事部门 vlan 20 的两台主机 PC3 和 PC6 之间无法通信。

```
PC>ping 192.168.0.3
ping 192.168.0.3: 32 data bytes, Press Ctrl_C to break
From 192.168.0.1: Destination host unreachable
From 192.168.0.1: Destination host unreachable
From 192.168.0.1: Destination host unreachable
From 192.168.0.1: Destination host unreachable
From 192.168.0.1: Destination host unreachable
--- 192.168.0.3 ping statistics ---
 5 packet(s) transmitted
 0 packet(s) received
 100.00% packet loss
```

5. 配置 Trunk 接口

上述测试结果显示，同部门的 PC3 和 PC6 之间无法通信。这是由于 PC3 和 PC6 虽然同属于 vlan 20，但是在交换机与交换机之间相连接口上并没有相应 VLAN 信息，不能够识别和发送跨越交换机的 VLAN 报文，此时 VLAN 只具有在每台交换机上的本地意义，无法实现相同 VLAN 的跨交换机通信。也可以理解为 SW1 和 SW2 之间连接接口 Etherent 0/0/4 分别属于默认 vlan 1，只允许 vlan 1 的报文通过。

为了让交换机间能够识别和发送跨越交换机的 VLAN 报文，需将交换机间连接的接口配置成为 Trunk 接口。配置时要明确被允许通过的 VLAN，实现对 VLAN 流量传输的控制。

在 SW1 上使用 **link-type** 命令配置 Ethernet 0/0/4 为 Trunk 接口，使用 **port trunk allow-pass** 命令配置允许 vlan 10 和 vlan 20 通过。

```
[SW1]interface Ethernet 0/0/4
[SW1-Ethernet0/0/4]port link-type trunk
[SW1-Ethernet0/0/4]port trunk allow-pass vlan 10 20
```

在 SW2 上配置 Ethernet 0/0/4 为 Trunk 接口，允许所有 vlan 通过。

```
[SW2]interface Ethernet 0/0/4
[SW2-Ethernet0/0/4]port link-type trunk
[SW2-Ethernet0/0/4]port trunk allow-pass vlan all
```

配置完成后可以使用 display port vlan 命令来检查 Trunk 的配置情况。

```
[SW1]display port vlan
Port                    Link Type    PVID   Trunk vlan List
-------------------------------------------------------------
Ethernet0/0/1           access       10     -
Ethernet0/0/2           access       10     -
Ethernet0/0/3           access       20     -
Ethernet0/0/4           trunk        1      1 10 20

[SW2]display port vlan
Port                    Link Type    PVID   Trunk vlan List
-------------------------------------------------------------
Ethernet0/0/1           access       30     -
Ethernet0/0/2           access       40     -
Ethernet0/0/3           access       20     -
Ethernet0/0/4           trunk        1      1-4094
```

可以观察 SW1 的 Etherent 0/0/4 被配置为 Turnk 接口，并允许 vlan 1、vlan 10 和 vlan 20 流量通过；SW2 的 Etherent 0/0/4 已经被配置为 Trunk 接口，并允许所有 VLAN 流量通过（vlan 1～vlan 4096）。

再次验证不同交换机上的相同部门的 PC 之间连通性，测试 PC3 和 PC6 之间连通性。

```
PC>ping 192.168.0.6
ping 192.168.0.6: 32 data bytes, Press Ctrl_C to break
From 192.168.0.6: bytes=32 seq=1 ttl=128 time=62 ms
From 192.168.0.6: bytes=32 seq=2 ttl=128 time=46 ms
From 192.168.0.6: bytes=32 seq=3 ttl=128 time=62 ms
From 192.168.0.6: bytes=32 seq=4 ttl=128 time=46 ms
From 192.168.0.6: bytes=32 seq=5 ttl=128 time=62 ms
--- 192.168.0.6 ping statistics ---
 5 packet(s) transmitted
 5 packet(s) received
 0.00% packet loss
 round-trip min/avg/max = 46/55/62 ms
```

4.4 Eth-Trunk 链路聚合配置

一、原理概述

以太网链路聚合 Eth-Trunk 简称链路聚合，它通过将多条以太网物理链路捆绑在一起成为一条逻辑链路，从而实现增加链路带宽的目的。同时，这些捆绑在一起的链路通过相互间的动态备份，可以有效地提高链路的可靠性。链路聚合主要有以下三个优势。

（1）提高带宽：聚合链路将多条链路虚拟成一条链路，带宽为多条链路带宽之和。

（2）提高可靠性：设备之间有一条链路发生故障不会影响设备之间的通信，除非聚合链路中的物理链路都断开，否则聚合链路不会失效。

（3）负载分担：在一个链路聚合组内，可以实现在各成员活动链路上的负载分担。

聚合链路的配置模式分为以下两种。

（1）手动负载模式：Eth-Trunk 的建立、成员接口的加入由手工配置，没有 LACP 的参与。当需要在两个直连设备之间提供一个较大的链路带宽而设备又不支持 LACP 协议时，可以使用手工模式。手工模式可以实现增加带宽、提高可靠性和负载分担的目的。

（2）LACP 模式：链路聚合控制协议（link aggregation control protocol，LACP）为交换数据的设备提供一种标准的协商方式，以供设备根据自身配置自动形成聚合链路并启动聚合链路收发数据。聚合链路形成以后，LACP 负责维护链路状态，在聚合条件发生变化时，自动调整或解散链路聚合。

二、实验目的

1. 理解 Eth-Trunk 链路聚合的原理
2. 掌握手动配置 Eth-Trunk 链路聚合的方法
3. 掌握 LACP 配置 Eth-Trunk 链路聚合的方法

三、实验内容

本实验模拟企业网络场景。SW1 和 SW2 为企业核心交换机，PC1 和 PC2 分别为行政部和研发部的主机。SW1 与 SW2 之间仅一条 Trunk 链路，考虑 Trunk 线路的带宽和可靠性，增加两条物理链路连接，采用 Eth-Trunk 实现需求。

四、实验拓扑

Eth-Trunk 链路聚合配置的拓扑如图 4.6 所示。

第 4 章　交换机配置

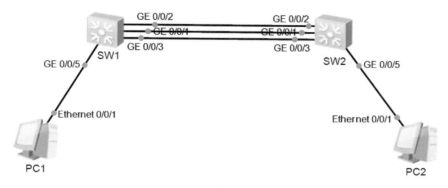

图 4.6　Eth-Trunk 链路聚合配置拓扑

五、实验地址分配

本实验的 IP 地址分配如表 4.4 所示。

表 4.4　IP 地址分配表

设备	接口	IP 地址	子网掩码	默认网关
PC1	Ethernet 0/0/1	192.168.9.16	255.255.255.0	N/A
PC2	Ethernet 0/0/1	192.168.9.18	255.255.255.0	N/A

六、实验步骤

1. 基本配置

根据实验地址分配表进行相应的基本配置。

1) PC1 的基本配置

PC1 的基本配置如图 4.7 所示。

图 4.7　PC1 的基本配置

2) PC2 的基本配置

PC2 的基本配置如图 4.8 所示。

图 4.8　PC2 的基本配置

3）测试 PC1 与 PC2 之间的连通性

在 PC1 上使用 ping 命令测试 PC1 与 PC2 之间的连通性。

```
PC>ping 192.168.9.18
ping 192.168.9.18: 32 data bytes, Press Ctrl_C to break
From 192.168.9.18: bytes=32 seq=1 ttl=128 time=63 ms
From 192.168.9.18: bytes=32 seq=2 ttl=128 time=47 ms
From 192.168.9.18: bytes=32 seq=3 ttl=128 time=78 ms
From 192.168.9.18: bytes=32 seq=4 ttl=128 time=47 ms
From 192.168.9.18: bytes=32 seq=5 ttl=128 time=62 ms
--- 192.168.9.18 ping statistics ---
  5 packet(s) transmitted
  5 packet(s) received
  0.00% packet loss
  round-trip min/avg/max = 47/59/78 ms
```

结果显示，PC1 与 PC2 之间通信正常。

4）查看交换机的 STP 状态

在 SW1 上使用 display stp brief 命令查看 STP 状态信息。

```
<Huawei>system-view
[Huawei]sysname SW1
[SW1]display stp brief
 MSTID   Port                   Role  STP State    Protection
   0     GigabitEthernet0/0/1   ROOT  FORWARDING   NONE
   0     GigabitEthernet0/0/2   ALTE  DISCARDING   NONE
   0     GigabitEthernet0/0/3   ALTE  DISCARDING   NONE
   0     GigabitEthernet0/0/5   DESI  FORWARDING   NONE
```

结果显示，SW1 的 GE 0/0/2 和 GE 0/0/3 接口处于丢弃状态。

在 SW2 上使用 display stp brief 命令查看 STP 状态信息。

```
<Huawei>system-view
[Huawei]sysname SW2
[SW2]display stp brief
 MSTID  Port                       Role  STP State   Protection
   0    GigabitEthernet0/0/1       DESI  FORWARDING  NONE
   0    GigabitEthernet0/0/2       DESI  FORWARDING  NONE
   0    GigabitEthernet0/0/3       DESI  FORWARDING  NONE
   0    GigabitEthernet0/0/5       DESI  FORWARDING  NONE
```

结果显示，SW2 的所有接口处于转发状态。

综上，如果要增加 SW1 和 SW2 之间的带宽，显然单靠增加链路条数是不够的，因为生成树会阻塞多余的接口，使得 SW1 与 SW2 之间的数据仅通过 GE 0/0/1 接口传输。

2. 手动配置 Eth-Trunk 链路聚合

1) 创建 Eth-Trunk 接口

在 SW1 上配置链路聚合，创建 Eth-Trunk 1 接口，并使用 **mode manual load-balance** 命令指定为手工模式。

```
[SW1]interface Eth-Trunk 1
[SW1-Eth-Trunk1]mode manual load-balance
```

在 SW2 上配置链路聚合，创建 Eth-Trunk 1 接口，并使用 **mode manual load-balance** 命令指定为手工模式。

```
[SW2]interface Eth-Trunk 1
[SW2-Eth-Trunk1]mode manual load-balance
```

2) 将相关接口加入 Eth-Trunk 接口

在 SW1 上将 GE 0/0/1、GE 0/0/2 和 GE 0/0/3 分别加入 Eth-Trunk 1 接口。

```
[SW1]interface GigabitEthernet 0/0/1
[SW1-GigabitEthernet0/0/1]eth-trunk 1
[SW1-GigabitEthernet0/0/1]interface GigabitEthernet 0/0/2
[SW1-GigabitEthernet0/0/2]eth-trunk 1
[SW1-GigabitEthernet0/0/2]interface GigabitEthernet 0/0/3
[SW1-GigabitEthernet0/0/3]eth-trunk 1
```

在 SW2 上将 GE 0/0/1、GE 0/0/2 和 GE 0/0/3 分别加入 Eth-Trunk 1 接口。

```
[SW2]interface GigabitEthernet 0/0/1
[SW2-GigabitEthernet0/0/1]eth-trunk 1
[SW2-GigabitEthernet0/0/1]interface GigabitEthernet 0/0/2
[SW2-GigabitEthernet0/0/2]eth-trunk 1
[SW2-GigabitEthernet0/0/2]interface GigabitEthernet 0/0/3
[SW2-GigabitEthernet0/0/3]eth-trunk 1
```

在 SW1 上使用 display eth-trunk 1 命令显示 Eth-Trunk 1 接口状态。

```
[SW1]display eth-trunk 1
Eth-Trunk1's state information is:
WorkingMode: NORMAL        Hash arithmetic: According to SIP-XOR-DIP
Least Active-linknumber: 1 Max Bandwidth-affected-linknumber: 8
Operate status: up         Number Of Up Port In Trunk: 3
--------------------------------------------------------------------
PortName                  Status      Weight
GigabitEthernet0/0/1       Up          1
GigabitEthernet0/0/2       Up          1
GigabitEthernet0/0/3       Up          1
```

结果显示，SW1 的工作模式为 NORMAL（手工模式），GE 0/0/1、GE 0/0/2 和 GE 0/0/3 接口已经添加到 Eth-Trunk 1 接口中，状态为 Up。

在 SW2 上使用 display eth-trunk 1 命令显示 Eth-Trunk 1 接口状态。

```
[SW2]display eth-trunk 1
Eth-Trunk1's state information is:
WorkingMode: NORMAL        Hash arithmetic: According to SIP-XOR-DIP
Least Active-linknumber: 1 Max Bandwidth-affected-linknumber: 8
Operate status: up         Number Of Up Port In Trunk: 3
--------------------------------------------------------------------
PortName                  Status      Weight
GigabitEthernet0/0/1       Up          1
GigabitEthernet0/0/2       Up          1
GigabitEthernet0/0/3       Up          1
```

结果显示，SW2 的工作模式为 NORMAL（手工模式），GE 0/0/1、GE 0/0/2 和 GE 0/0/3 已经添加到 Eth-Trunk 1 中，状态为 Up。

3) 查看 Eth-Trunk 接口信息

在 SW1 上使用 display interface eth-trunk 1 命令查看 SW1 的 Eth-Trunk 接口信息。

```
[SW1]display interface eth-trunk 1
Eth-Trunk1 current state : UP
Line protocol current state : UP
Description:
Switch Port,PVID:1,Hash arithmetic:According to SIP-XOR-DIP,Maximal BW:3G,
Current BW: 3G, The Maximum Frame Length is 9216
 IP Sending Frames'Format is PKTFMT_ETHNT_2,Hardware address is
4c1f-cca6-5ce6
 Current system time: 2022-10-02 16:15:33-08:00
   Input bandwidth utilization  :    0%
   Output bandwidth utilization :    0%
--------------------------------------------------
PortName                    Status    Weight
--------------------------------------------------
GigabitEthernet0/0/1         UP        1
GigabitEthernet0/0/2         UP        1
GigabitEthernet0/0/3         UP        1
--------------------------------------------------
The Number of Ports in Trunk : 3
The Number of UP Ports in Trunk : 3
```

结果显示，目前 Eth-Trunk 接口的总带宽为 GE 0/0/1、GE 0/0/2 和 GE 0/0/3 接口带宽之和，即 3 Gb/s。

在 SW1 上使用 **display stp brief** 命令查看生成树状态。

```
[SW1]display stp brief
 MSTID   Port                    Role  STP State   Protection
   0     GigabitEthernet0/0/5    DESI  FORWARDING  NONE
   0     Eth-Trunk1              ROOT  FORWARDING  NONE
```

结果显示，SW1 的 3 个接口被捆绑成一个 Eth-Trunk 接口，该接口处于转发状态。

4）测试 PC1 与 PC2 之间的连通性

在 PC1 上使用 **ping** 命令持续测试与 PC2 的连通性。同时将 SW2 的 GE 0/0/1 或 GE 0/0/2 或 GE 0/0/3 关闭模拟故障发生（关闭其中一个）。

```
PC>ping 192.168.9.18
ping 192.168.9.18: 32 data bytes, Press Ctrl_C to break
From 192.168.9.18: bytes=32 seq=1 ttl=128 time=109 ms
From 192.168.9.18: bytes=32 seq=2 ttl=128 time=109 ms
From 192.168.9.18: bytes=32 seq=3 ttl=128 time=110 ms
From 192.168.9.18: bytes=32 seq=4 ttl=128 time=93 ms
From 192.168.9.18: bytes=32 seq=5 ttl=128 time=110 ms
Request timeout!
```

```
Request timeout!
……
From 192.168.9.18: bytes=32 seq=33 ttl=128 time=78 ms
From 192.168.9.18: bytes=32 seq=34 ttl=128 time=94 ms
From 192.168.9.18: bytes=32 seq=35 ttl=128 time=93 ms
From 192.168.9.18: bytes=32 seq=36 ttl=128 time=63 ms
```

结果显示，关闭 Eth-Trunk 1 中的任意一个接口，照常正常通信。说明链路聚合除了可以增加带宽，还可以容错。

3. LACP 配置 Eth-Trunk 链路聚合

1) 删除加入 Eth-Trunk 接口下的物理接口

在 SW1 上使用 undo eth-trunk 命令删除加入 Eth-Trunk 接口的 GE 0/0/1 或 GE 0/0/2 或 GE 0/0/3。

```
[SW1]interface GigabitEthernet 0/0/1
[SW1-GigabitEthernet0/0/1]undo eth-trunk
[SW1-GigabitEthernet0/0/1]interface GigabitEthernet 0/0/2
[SW1-GigabitEthernet0/0/2]undo eth-trunk
[SW1-GigabitEthernet0/0/2]interface GigabitEthernet 0/0/3
[SW1-GigabitEthernet0/0/3]undo eth-trunk
```

在 SW2 上使用 undo eth-trunk 命令删除加入 Eth-Trunk 接口的 GE 0/0/1 或 GE 0/0/2 或 GE 0/0/3。

```
[SW2]interface GigabitEthernet 0/0/1
[SW2-GigabitEthernet0/0/1]undo eth-trunk
[SW2-GigabitEthernet0/0/1]interface GigabitEthernet 0/0/2
[SW2-GigabitEthernet0/0/2]undo eth-trunk
[SW2-GigabitEthernet0/0/2]interface GigabitEthernet 0/0/3
[SW2-GigabitEthernet0/0/3]undo eth-trunk
```

2) 将 Eth-Trunk 的工作模式改为 LACP 模式

在 SW1 上使用 mode lacp-static 命令将 Eth-Trunk 接口的工作模式改为 LACP 模式，并将 GE 0/0/1 或 GE 0/0/2 或 GE 0/0/3 接口分别加入 Eth-Trunk 接口。

```
[SW1]interface eth-trunk 1
[SW1-Eth-Trunk1]mode lacp-static
[SW1-Eth-Trunk1]lacp preempt enable
[SW1-Eth-Trunk1]interface Gigabitethernet 0/0/1
[SW1-GigabitEthernet0/0/1]eth-trunk 1
```

```
[SW1-GigabitEthernet0/0/1]interface Gigabitethernet 0/0/2
[SW1-GigabitEthernet0/0/2]eth-trunk 1
[SW1-GigabitEthernet0/0/2]interface Gigabitethernet 0/0/3
[SW1-GigabitEthernet0/0/3]eth-trunk 1
```

在 SW2 上使用 mode lacp-static 命令将 Eth-Trunk 接口的工作模式改为 LACP 模式，并将 GE 0/0/1 或 GE 0/0/2 或 GE 0/0/3 接口分别加入 Eth-Trunk 接口。

```
[SW2]interface eth-trunk 1
[SW2-Eth-Trunk1]mode lacp-static
[SW2-Eth-Trunk1]lacp preempt enable
[SW2-Eth-Trunk1]interface Gigabitethernet 0/0/1
[SW2-GigabitEthernet0/0/1]eth-trunk 1
[SW2-GigabitEthernet0/0/1]interface Gigabitethernet 0/0/2
[SW2-GigabitEthernet0/0/2]eth-trunk 1
[SW2-GigabitEthernet0/0/2]interface Gigabitethernet 0/0/3
[SW2-GigabitEthernet0/0/3]eth-trunk 1
```

3）查看 Eth-Trunk 接口状态

在 SW1 上使用 display eth-trunk 1 命令查看 Eth-Trunk 接口状态。

```
[SW1]display eth-trunk 1
Eth-Trunk1's state information is:
Local:
LAG ID: 1                    WorkingMode: STATIC
Preempt Delay: Disabled      Hash arithmetic: According to SIP-XOR-DIP
System Priority: 32768       System ID: 4c1f-cca6-5ce6
Least Active-linknumber: 1   Max Active-linknumber: 8
Operate status: up           Number Of Up Port In Trunk: 2
--------------------------------------------------------------------------
ActorPortName         Status    PortType PortPri PortNo PortKey PortState Weight
GigabitEthernet0/0/1  selected  1GE      32768   2      305     10111100  1
GigabitEthernet0/0/2  Selected  1GE      32768   3      305     10111100  1
GigabitEthernet0/0/3  Selected  1GE      32768   4      305     10111100  1
Partner:
--------------------------------------------------------------------------
ActorPortName         SysPri  SystemID        PortPri PortNo PortKey PortState
GigabitEthernet0/0/1  32768   4c1f-cc41-050c  32768   2      305     10111100
GigabitEthernet0/0/2  32768   4c1f-cc41-050c  32768   3      305     10111100
GigabitEthernet0/0/3  32768   4c1f-cc41-050c  32768   4      305     10111100
```

在 SW2 上使用 display eth-trunk 1 命令查看 Eth-Trunk 接口状态。

```
[SW2]display eth-trunk 1
Eth-Trunk1's state information is:
Local:
LAG ID: 1                      WorkingMode: STATIC
Preempt Delay: Disabled        Hash arithmetic: According to SIP-XOR-DIP
System Priority: 32768         System ID: 4c1f-cc41-050c
Least Active-linknumber: 1  Max Active-linknumber: 8
Operate status: up             Number Of Up Port In Trunk: 2
--------------------------------------------------------------------------
ActorPortName         Status   PortType PortPri PortNo PortKey PortState Weight
GigabitEthernet0/0/1 selected  1GE      32768   2      305     10100010  1
GigabitEthernet0/0/2 Selected  1GE      32768   3      305     10111100  1
GigabitEthernet0/0/3 Selected  1GE      32768   4      305     10111100  1
Partner:
--------------------------------------------------------------------------
ActorPortName        SysPri   SystemID        PortPri PortNo PortKey PortState
GigabitEthernet0/0/1 32768    4c1f-cca6-5ce6  32768   2      305     10111100
GigabitEthernet0/0/2 32768    4c1f-cca6-5ce6  32768   3      305     10111100
GigabitEthernet0/0/3 32768    4c1f-cca6-5ce6  32768   4      305     10111100
```

结果显示，SW1 和 SW2 上的 GE 0/0/1、GE 0/0/2 和 GE 0/0/3 处于活动状态（Selected）。

将 SW1 的系统优先级从默认的 32768 改为 200，使其成为主动端（值越低优先级越高），并按主动端设备的接口来选择活动端口。两端设备选出主动端后，两端都会以主动端的接口优先级来选择活动接口。两端设备选择了一致的活动接口，活动链路才可以建立起来。设置这些活动链路以负载分担的方式转发数据。

```
[SW1]lacp priority 200
[SW1]display eth-trunk 1
Eth-Trunk1's state information is:
Local:
LAG ID: 1                      WorkingMode: STATIC
Preempt Delay Time: 30         Hash arithmetic: According to SIP-XOR-DIP
System Priority: 200           System ID: 4c1f-cca6-5ce6
Least Active-linknumber: 1  Max Active-linknumber: 8
……
```

结果显示，SW1 的 LACP 系统优先级改为 200，因 SW2 未修改，仍为默认值。

在 SW1 上使用 max active-linknumber 2 配置活动接口阈值为 2，并配置 GE 0/0/1 和 GE 0/0/2 接口优先级确定活动链路。

```
[SW1]interface eth-trunk 1
[SW1-Eth-Trunk1]max active-linknumber 2
```

```
[SW1-Eth-Trunk1]quit
[SW1]interface GigabitEthernet 0/0/1
[SW1-GigabitEthernet0/0/1]lacp priority 200
[SW1-GigabitEthernet0/0/1]interface GigabitEthernet 0/0/2
[SW1-GigabitEthernet0/0/2]lacp priority 200
```

配置完成后，在 SW1 上查看 Eth-Trunk 1 接口状态。

```
[SW1]display eth-trunk 1
Eth-Trunk1's state information is:
Local:
LAG ID: 1                     WorkingMode: STATIC
Preempt Delay Time: 30        Hash arithmetic: According to SIP-XOR-DIP
System Priority: 200          System ID: 4c1f-cca6-5ce6
Least Active-linknumber: 1    Max Active-linknumber: 2
Operate status: up            Number Of Up Port In Trunk: 2
--------------------------------------------------------------------
ActorPortName            Status    PortType PortPri PortNo PortKey PortState Weight
GigabitEthernet0/0/1     Selected  1GE      200     2      305     10111100  1
GigabitEthernet0/0/2     Selected  1GE      200     3      305     10111100  1
GigabitEthernet0/0/3     Unselect  1GE      32768   4      305     10100000  1
Partner:
……
```

结果显示，由于接口的阈值改为 2（默认活动接口最大阈值为 8），该 Eth-Trunk 接口下将只有两个成员处于活动状态，并且具有负载分担能力。而 GE 0/0/3 接口已处于不活动状态（Unselect），该链路作为备份链路。当活动链路出现故障时，备份链路将会替代故障链路，保持数据的可靠性。

将 SW1 的 GE 0/0/1 接口关闭，验证 Eth-Trunk 链路聚合信息。

```
[SW1]interface GigabitEthernet 0/0/1
[SW1-GigabitEthernet0/0/1]shutdown
[SW1-GigabitEthernet0/0/1]quit
[SW1]display eth-trunk 1
Eth-Trunk1's state information is:
Local:
LAG ID: 1                     WorkingMode: STATIC
Preempt Delay Time: 30        Hash arithmetic: According to SIP-XOR-DIP
System Priority: 200          System ID: 4c1f-cca6-5ce6
Least Active-linknumber: 1    Max Active-linknumber: 2
Operate status: up            Number Of Up Port In Trunk: 2
--------------------------------------------------------------------
ActorPortName            Status    PortType PortPri PortNo PortKey PortState Weight
GigabitEthernet0/0/1     Unselect  1GE      200     2      305     10100010  1
```

```
GigabitEthernet0/0/2  Selected  1GE      200     3   305  10111100  1
GigabitEthernet0/0/3  Selected  1GE      32768   4   305  10111100  1
Partner:
--------------------------------------------------------------------
ActorPortName         SysPri    SystemID         PortPri PortNo PortKey PortState
GigabitEthernet0/0/1  0         0000-0000-0000   0       0      0       10100011
GigabitEthernet0/0/2  32768     4c1f-cc41-050c   32768   3      305     10111100
GigabitEthernet0/0/3  32768     4c1f-cc41-050c   32768   4      305     10111100
```

结果显示，SW1 的 GE 0/0/1 接口已经处于不活动状态，而 GE 0/0/3 接口为活动状态。如果 SW1 的 GE 0/0/1 接口开启后，又会恢复为活动状态，GE 0/0/3 则为不活动状态。

在 SW2 上查看 Eth-Trunk 1 接口状态。

```
[SW2]display eth-trunk 1
Eth-Trunk1's state information is:
Local:
LAG ID: 1                    WorkingMode: STATIC
Preempt Delay Time: 30       Hash arithmetic: According to SIP-XOR-DIP
System Priority: 32768       System ID: 4c1f-cc41-050c
Least Active-linknumber: 1   Max Active-linknumber: 8
Operate status: up           Number Of Up Port In Trunk: 2
--------------------------------------------------------------------
ActorPortName         Status    PortType PortPri PortNo PortKey PortState Weight
GigabitEthernet0/0/1  Unselect  1GE      32768   2      305     10100010  1
GigabitEthernet0/0/2  Selected  1GE      32768   3      305     10111100  1
GigabitEthernet0/0/3  Selected  1GE      32768   4      305     10111100  1
Partner:
--------------------------------------------------------------------
ActorPortName         SysPri    SystemID         PortPri PortNo PortKey PortState
GigabitEthernet0/0/1  0         0000-0000-0000   0       0      0       10100011
GigabitEthernet0/0/2  200       4c1f-cca6-5ce6   200     3      305     10111100
GigabitEthernet0/0/3  200       4c1f-cca6-5ce6   32768   4      305     10111100
```

结果显示，SW2 随主动端发生变化。SW2 的 LACP 系统优先级没修改，仍为默认值。

第 5 章 路由器配置

5.1 静态路由配置

一、原理概述

静态路由是指用户或网络管理员手工配置的路由信息。静态路由配置简单，比较适合小型、简单的互联环境。静态路由不适合大型、复杂的互联环境，因为当网络拓扑发生变化时，网络管理员要做大量的调整，且无法自动检测错误发生，不易排错。

默认路由是一种特殊的静态路由，当路由表中的数据包目的地址没有匹配的表项时，数据包将根据默认路由表项进行转发。默认路由可以大大简化路由器配置，减轻网络管理员的工作负担。

二、实验目的

1. 掌握配置和测试静态路由（指定接口或下一跳 IP 地址）的方法
2. 掌握配置和测试默认路由的方法
3. 掌握在简单网络中部署静态路由时的故障排除方法
4. 掌握简单的网络优化方法

三、实验内容

某公司拥有 3 台路由器构建的互联网，R1、R2 和 R3 各自连接一台主机，现在要求能够实现主机之间的正常通信。本实验通过配置静态路由和默认路由来实现。

四、实验拓扑

静态路由配置拓扑如图 5.1 所示。

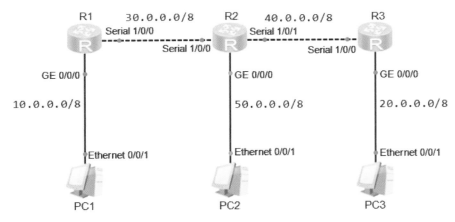

图 5.1 静态路由配置拓扑

五、实验地址分配

本实验 IP 地址分配如表 5.1 所示。

表 5.1 IP 地址分配

设备	接口	IP 地址	子网掩码	默认网关
PC1	Ethernet 0/0/1	10.0.0.2	255.0.0.0	10.0.0.1
R1（AR2220）	GE 0/0/0	10.0.0.1	255.0.0.0	
	Serial 1/0/0	30.0.0.1	255.0.0.0	
R2（AR2220）	Serial 1/0/0	30.0.0.2	255.0.0.0	
	Serial 1/0/1	40.0.0.1	255.0.0.0	
	GE 0/0/0	50.0.0.1	255.0.0.0	
PC2	Ethernet 0/0/1	50.0.0.2	255.0.0.0	50.0.0.1
R3（AR2220）	Serial 1/0/0	40.0.0.2	255.0.0.0	
	GE 0/0/0	20.0.0.1	255.0.0.0	
PC3	Ethernet 0/0/1	20.0.0.2	255.0.0.0	20.0.0.1

六、实验步骤

1. 拓扑

拓扑设计时，为 3 个路由器分别添加 Serial 接口卡，在各个路由器右键快捷菜单中单击"配置"，在 eNSP 支持的接口卡中选取 2SA 接口卡，拖放至上方的空位置，如图 5.2 所示。

第 5 章 路由器配置

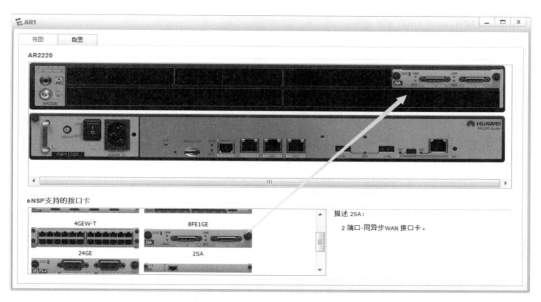

图 5.2　添加 Serial 接口卡

2. 基本配置

1) PC 的基本配置

根据实验地址分配表进行相应的基本配置。

PC1 的基本配置如图 5.3 所示。

图 5.3　PC1 的基本配置

PC2 的基本配置如图 5.4 所示。

图 5.4　PC2 的基本配置

PC3 的基本配置如图 5.5 所示。

```
IPv4 配置
●静态        ○DHCP                    □自动获取 DNS 服务器地址
IP 地址：    20 . 0 . 0 . 2            DNS1:    0 . 0 . 0 . 0
子网掩码：   255 . 0 . 0 . 0           DNS2:    0 . 0 . 0 . 0
网关：       20 . 0 . 0 . 1
```

图 5.5 PC3 的基本配置

2）路由器的基本配置

根据实验地址分配表完成路由器 R1、R2 和 R3 的基本配置，包括命名、各接口 IP 地址的配置。

(1) R1 的配置。

```
<Huawei>system-view
[Huawei]sysname R1
[R1]interface GigabitEthernet 0/0/0
[R1-GigabitEthernet0/0/0]ip address 10.0.0.1 255.0.0.0
[R1-GigabitEthernet0/0/0]interface serial 1/0/0
[R1-Serial1/0/0]ip address 30.0.0.1 255.0.0.0
```

(2) R2 的配置。

```
<Huawei>system-view
[Huawei]sysname R2
[R2]interface serial 1/0/0
[R2-Serial1/0/0]ip address 30.0.0.2 255.0.0.0
[R2-Serial1/0/0]interface serial 1/0/1
[R2-Serial1/0/1]ip address 40.0.0.1 255.0.0.0
[R2-Serial1/0/1]interface Gigabitethernet 0/0/0
[R2-GigabitEthernet0/0/0]ip address 50.0.0.1 255.0.0.0
```

(3) R3 的配置。

```
<Huawei>system-view
[Huawei]sysname R3
[R3]interface serial 1/0/0
[R3-Serial1/0/0]ip address 40.0.0.2 255.0.0.0
[R3-Serial1/0/0]interface GigabitEthernet 0/0/0
```

```
[R3-GigabitEthernet0/0/0]ip address 20.0.0.1 255.0.0.0
```

3) 测试结果

(1) 测试各直连线的连通性。

在 R1 上使用 ping 命令测试直连线的连通性，其余直连网段的连通性测试省略。

```
<R1>ping 10.0.0.2    //测试与PC1的连通性
 ping 10.0.0.2: 56  data bytes, press CTRL_C to break
   Reply from 10.0.0.2: bytes=56 Sequence=1 ttl=128 time=50 ms
   Reply from 10.0.0.2: bytes=56 Sequence=2 ttl=128 time=20 ms
   Reply from 10.0.0.2: bytes=56 Sequence=3 ttl=128 time=20 ms
   Reply from 10.0.0.2: bytes=56 Sequence=4 ttl=128 time=20 ms
   Reply from 10.0.0.2: bytes=56 Sequence=5 ttl=128 time=20 ms
 --- 10.0.0.2 ping statistics ---
   5 packet(s) transmitted
   5 packet(s) received
   0.00% packet loss
   round-trip min/avg/max = 20/26/50 ms
<R1>ping 30.0.0.2    //测试与R2的连通性
 ping 30.0.0.2: 56  data bytes, press CTRL_C to break
   Reply from 30.0.0.2: bytes=56 Sequence=1 ttl=255 time=60 ms
   Reply from 30.0.0.2: bytes=56 Sequence=2 ttl=255 time=30 ms
   Reply from 30.0.0.2: bytes=56 Sequence=3 ttl=255 time=20 ms
   Reply from 30.0.0.2: bytes=56 Sequence=4 ttl=255 time=20 ms
   Reply from 30.0.0.2: bytes=56 Sequence=5 ttl=255 time=10 ms
 --- 30.0.0.2 ping statistics ---
   5 packet(s) transmitted
   5 packet(s) received
   0.00% packet loss
   round-trip min/avg/max = 10/28/60 ms
```

(2) 测试跨网段的连通性。

在 PC1 上使用 ping 命令测试跨网段的连通性，其余跨网段的连通性测试省略。

```
PC>ping 20.0.0.2    //测试与PC3的连通性
ping 20.0.0.2: 32 data bytes, Press Ctrl_C to break
Request timeout!
Request timeout!
Request timeout!
Request timeout!
Request timeout!
--- 20.0.0.2 ping statistics ---
  5 packet(s) transmitted
  0 packet(s) received
```

```
    100.00% packet loss
PC>ping 50.0.0.2    //测试与PC2的连通性
ping 50.0.0.2: 32 data bytes, Press Ctrl_C to break
Request timeout!
Request timeout!
Request timeout!
Request timeout!
Request timeout!
--- 50.0.0.2 ping statistics ---
  5 packet(s) transmitted
  0 packet(s) received
  100.00% packet loss
```

结果显示无法连通，这时需要思考是什么问题导致了它们之间无法通信。

(3) 检查路由表。

在 R1 上使用 **display ip routing-table** 检查路由表。

```
<R1>display ip routing-table   //查看R1的路由表
Route Flags: R - relay, D - download to fib
------------------------------------------------------------------------
Routing Tables: Public
    Destinations : 11      Routes : 11
Destination/Mask    Proto  Pre  Cost    Flags NextHop    Interface
10.0.0.0/8          Direct 0    0       D     10.0.0.1   GigabitEthernet0/0/0
30.0.0.0/8          Direct 0    0       D     30.0.0.1   Serial1/0/0
......
InLoopBack0
```

R1 的路由表上没有任何关于 PC2 和 PC3 所在网段的信息（50.0.0.0/8 和 20.0.0.0/8）。可以使用相同的方法查看 R2 和 R3 的路由表。R2 上没有任何关于 PC1 和 PC2 所在网段的信息，R3 上没有任何关于 PC1 和 PC3 的信息，验证了初始状态下各路由器的路由表上仅包括了与自身直接相连的网段的路由信息。

4) 配置路由表

下面分别在 R1 和 R2 上进行静态路由配置，在 R3 上进行默认路由配置，然后测试结果。

(1) 配置静态路由表。

在 R1 上配置间接到达网段的静态路由，即目的网络地址分别为 20.0.0.0/8、40.0.0.0/8 和 50.0.0.2/8。对于 R1 而言，要发送数据到主机 PC2 和 PC3，都须先发送给 R2，所以 R2 即为 R1 的下一跳路由器，R2 与 R1 所在的直连链路上的物理接口的 IP 地址即为下一跳 IP 地址，即下一跳 IP 地址为 30.0.0.2。在 R1 上使用 **ip route-static** 命令配置静态路由表。

```
[R1]ip route-static 20.0.0.0 255.0.0.0 30.0.0.2
[R1]ip route-static 40.0.0.0 255.0.0.0 30.0.0.2
[R1]ip route-static 50.0.0.0 255.0.0.0 30.0.0.2
```

配置完成后，查看 R1 上的路由表，显示 R1 到 PC2 和 PC3 所在网段的路由信息。

```
<R1>display ip routing-table
Route Flags: R - relay, D - download to fib
------------------------------------------------------------------
Routing Tables: Public
        Destinations : 13        Routes : 13
Destination/Mask    Proto   Pre  Cost   Flags   NextHop      Interface
  ......
  20.0.0.0/8        Static  60   0      RD      30.0.0.2     Serial1/0/0
  40.0.0.0/8        Static  60   0      RD      30.0.0.2     Serial1/0/0
  50.0.0.0/8        Static  60   0      RD      30.0.0.2     Serial1/0/0
  ......
```

同样，在 R2 上使用 ip route-static 命令配置静态路由表。两条路由条目分别采用指定下一跳 IP 地址的方法和指定接口的方法配置。

```
[R2]ip route-static 10.0.0.0 255.0.0.0 30.0.0.1
[R2]ip route-static 20.0.0.0 255.0.0.0 Serial 1/0/1
```

配置完成后，查看 R2 上的路由表，显示 R2 到 PC1 和 PC2 所在网段的路由信息。

```
<R2>display ip routing-table
Route Flags: R - relay, D - download to fib
------------------------------------------------------------------
Routing Tables: Public
        Destinations : 17        Routes : 17
Destination/Mask    Proto   Pre  Cost   Flags   NextHop      Interface
  10.0.0.0/8        Static  60   0      RD      30.0.0.1     Serial1/0/0
  20.0.0.0/8        Static  60   0      RD      40.0.0.2     Serial1/0/1
  ......
```

此时，在主机 PC1 上使用 ping 命令测试与 PC3 的连通性，结果可以连通；测试与 PC2 的连通性，结果无法连通。在 PC1 的 Ethernet 0/0/1 接口上进行抓包。右键单击 PC1 的 Ethernet 0/0/1 接口，选择"抓取"命令。然后在 PC1 上 ping 主机 PC2，可以观察如图 5.6 所示的结果。

```
No. Time        Source              Destination      Protocol Info
1 0.000000  HuaweiTe_BroadcastARP                    ARP      who has 10.0.0.1?  Tell 10.0.0.2
2 0.015000  HuaweiTe_HuaweiTe_ARP                    ARP      10.0.0.1 is at 00:e0:fc:87:11:4b
3 0.015000  10.0.0.2            20.0.0.2             ICMP     Echo (ping) request  (id=0xbdc7, seq(be/le)=1/256, ttl=128)
4 2.012000  10.0.0.2            20.0.0.2             ICMP     Echo (ping) request  (id=0xbfc7, seq(be/le)=2/512, ttl=128)
5 4.025000  10.0.0.2            20.0.0.2             ICMP     Echo (ping) request  (id=0xc1c7, seq(be/le)=3/768, ttl=128)
6 6.021000  10.0.0.2            20.0.0.2             ICMP     Echo (ping) request  (id=0xc3c7, seq(be/le)=4/1024, ttl=128)
7 8.018000  10.0.0.2            20.0.0.2             ICMP     Echo (ping) request  (id=0xc5c7, seq(be/le)=5/1280, ttl=128)
```

图 5.6　抓包结果

此时主机 PC1 仅发送了 ICMP 请求消息，并没有收到任何回应消息。原因在于现在仅仅实现了 PC1 与 R1 的通信，而 PC2 仍然无法发送数据给 PC1，所以同样需要配置 R3 的路由表。

(2) 配置默认路由表。

通过适当减少设备上的配置工作量，一方面能够帮助网络管理员在进行故障排除时更容易定位故障，且相对较少的配置量也能减少在配置时出错的可能，另一方面相对减少路由表的项目数，减少路由器查表需要的时间。

默认路由是一种特殊的静态路由，使用默认路由可以简化路由器上的配置。通过查看 R1 的路由表，R1 上存在两条手工配置的静态路由条目，它们的下一跳和发出接口都一致。路由器 R3 与 R1 类似，R3 到主机 PC1 和 PC3 所在网段的下一跳和发出接口都一致，因此，可以在 R3 配置默认路由，即目的网段和掩码为全 0，表示任何网络，下一跳为 40.0.0.1。配置完成后查看 R3 的路由表。

在 R3 上使用 **ip route-static** 命令配置默认路由，使用 **display ip routing-table** 命令查看路由表。

```
[R3]ip route-static 0.0.0.0 0.0.0.0 40.0.0.1
[R3]quit
<R3>display ip routing-table
Route Flags: R - relay, D - download to fib
------------------------------------------------------------
Routing Tables: Public
        Destinations : 12       Routes : 12
Destination/Mask    Proto   Pre   Cost    Flags   NextHop       Interface
    0.0.0.0/0       Static  60    0       RD      40.0.0.1      Serial1/0/0
    ......
```

(3) 测试结果。

在主机 PC1 上使用 **ping** 命令测试与主机 PC2 的连通性：

```
PC>ping 20.0.0.2
ping 20.0.0.2: 32 data bytes, Press Ctrl_C to break
From 20.0.0.2: bytes=32 seq=1 ttl=125 time=31 ms
From 20.0.0.2: bytes=32 seq=2 ttl=125 time=15 ms
From 20.0.0.2: bytes=32 seq=3 ttl=125 time=16 ms
From 20.0.0.2: bytes=32 seq=4 ttl=125 time=16 ms
```

```
From 20.0.0.2: bytes=32 seq=5 ttl=125 time=31 ms
--- 20.0.0.2 ping statistics ---
  5 packet(s) transmitted
  5 packet(s) received
  0.00% packet loss
  round-trip min/avg/max = 15/21/31 ms
```

在主机 PC1 上使用 ping 命令测试与主机 PC3 的连通性：

```
PC>ping 50.0.0.2
ping 50.0.0.2: 32 data bytes, Press Ctrl_C to break
From 50.0.0.2: bytes=32 seq=1 ttl=126 time=16 ms
From 50.0.0.2: bytes=32 seq=2 ttl=126 time=16 ms
From 50.0.0.2: bytes=32 seq=3 ttl=126 time=16 ms
From 50.0.0.2: bytes=32 seq=4 ttl=126 time=16 ms
From 50.0.0.2: bytes=32 seq=5 ttl=126 time=16 ms
--- 50.0.0.2 ping statistics ---
  5 packet(s) transmitted
  5 packet(s) received
  0.00% packet loss
  round-trip min/avg/max = 16/16/16 ms
```

结果表明，PC1 与 PC2、PC3 之间通信正常。

5.2 RIP 动态路由配置

一、原理描述

RIP（routing information protocol，路由协议）作为最早的距离矢量路由协议，也是最先得到广泛应用的一种路由协议，采用了 Bellman-Ford 算法，其最大的特点就是配置简单。

RIP 要求网络中每台路由器都要维护从自身到每一个目标网络的路由信息。RIP 使用"跳数"来衡量网络间的"距离"，从一台路由器到其直接相连的网络的跳数定义为 1，到间接相连的网络的距离定义为每经过一台路由器则距离加 1。RIP 允许路由器的最大跳数为 15，因此，跳数为 16 即为不可达。可见 RIP 只适用于小型互联网。

目前 RIP 有两个版本，即 RIPv1 和 RIPv2。RIPv2 针对 RIPv1 进行了扩充，能够携带更多的信息量，并增强了安全性能。RIPv1 和 RIPv2 都是基于 UDP 的协议，使用 UDP 的 520 端口收发数据包。

二、实验内容

某公司拥有 3 台路由器，因此可以采用 RIP 来完成网络的配置。本实验通过模拟简单的企业网络场景来描述 RIP 的基本配置，并介绍一些基本的 RIP 信息的命令使用方法。

三、实验目的

1. 理解 RIP 应用场景与基本原理
2. 掌握 RIPv1 和 RIPv2 的基本配置，了解 RIPv1 和 RIPv2 的区别
3. 掌握测试 RIP 网络的连通性的方法
4. 掌握使用 display 与 debug 命令测试 RIP 的方法

四、实验拓扑

RIP 动态路由配置拓扑如图 5.7 所示。

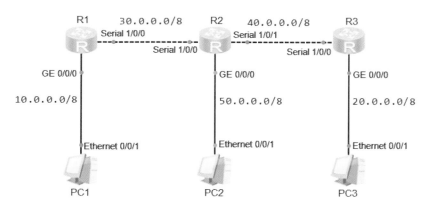

图 5.7　RIP 动态路由配置拓扑

五、实验地址分配

本实验 IP 地址分配如表 5.2 所示。

表 5.2　IP 地址分配

设备	接口	IP 地址	子网掩码	默认网关
PC1	Ethernet 0/0/1	10.0.0.2	255.0.0.0	10.0.0.1
R1（AR2220）	GE 0/0/0	10.0.0.1	255.0.0.0	
	Serial 1/0/0	30.0.0.1	255.0.0.0	
R2（AR2220）	Serial 1/0/0	30.0.0.2	255.0.0.0	
	Serial 1/0/1	40.0.0.1	255.0.0.0	
	GE 0/0/0	50.0.0.1	255.0.0.0	

续表

设备	接口	IP 地址	子网掩码	默认网关
PC2	Ethernet 0/0/1	50.0.0.2	255.0.0.0	50.0.0.1
R3（AR2220）	Serial 1/0/0	40.0.0.2	255.0.0.0	
	GE 0/0/0	20.0.0.1	255.0.0.0	
PC3	Ethernet 0/0/1	20.0.0.2	255.0.0.0	20.0.0.1

六、实验步骤

1. PC 的基本配置

根据实验地址分配表分别完成 PC1、PC2 和 PC3 的基本配置。

PC1 的基本配置如图 5.8 所示。

图 5.8　PC1 的基本配置

PC2 的基本配置如图 5.9 所示。

图 5.9　PC2 的基本配置

PC3 的基本配置如图 5.10 所示。

图 5.10　PC3 的基本配置

2. 路由器的基本配置

根据实验地址分配表分别完成 R1、R2 和 R3 的基本配置。

1) R1 的配置

```
<Huawei>system-view
[Huawei]sysname R1
[R1]interface GigabitEthernet 0/0/0
[R1-GigabitEthernet0/0/0]ip address 10.0.0.1 255.0.0.0
[R1-GigabitEthernet0/0/0]interface serial 1/0/0
[R1-Serial1/0/0]ip address 30.0.0.1 255.0.0.0
```

2) R2 的配置

```
<Huawei>system-view
[Huawei]sysname R2
[R2]interface serial 1/0/0
[R2-Serial1/0/0]ip address 30.0.0.2 255.0.0.0
[R2-Serial1/0/0]interface serial 1/0/1
[R2-Serial1/0/1]ip address 40.0.0.1 255.0.0.0
[R2-Serial1/0/1]interface Gigabitethernet 0/0/0
[R2-GigabitEthernet0/0/0]ip address 50.0.0.1 255.0.0.0
```

3) R3 的配置

```
<Huawei>system-view
[Huawei]sysname R3
[R3]interface serial 1/0/0
[R3-Serial1/0/0]ip address 40.0.0.2 255.0.0.0
[R3-Serial1/0/0]interface GigabitEthernet 0/0/0
[R3-GigabitEthernet0/0/0]ip address 20.0.0.1 255.0.0.0
```

在配置 RIP 之前，使用 ping 命令检测直连链路的连通性。例如，检测 R1 到 R2 之间直连链路的连通性。

```
<R1>ping -c 1 30.0.0.2
  ping 30.0.0.2: 56  data bytes, press CTRL_C to break
   Reply from 30.0.0.2: bytes=56 Sequence=1 ttl=255 time=20 ms
 --- 30.0.0.2 ping statistics ---
   1 packet(s) transmitted
   1 packet(s) received
   0.00% packet loss
```

第 5 章　路由器配置

```
round-trip min/avg/max = 20/20/20 ms
```

3. 使用 RIPv1 搭建网络

在路由器 R1、R2 和 R3 上配置 RIPv1。使用 rip 命令创建并开启协议进程，默认情况下进程号为 1。使用 network 命令对指定的网络接口使用 RIP 功能，注意必须是自然网段的网络地址（没有划分子网）。

```
[R1]rip
[R1-rip-1]network 10.0.0.0
[R1-rip-1]network 30.0.0.0

[R2]rip
[R2-rip-1]network 30.0.0.0
[R2-rip-1]network 40.0.0.0
[R2-rip-1]network 50.0.0.0

[R3]rip
[R3-rip-1]network 20.0.0.0
[R3-rip-1]network 40.0.0.0
```

配置完成后，使用 display ip routing-table 命令查看 R1 的路由表。

```
<R1>display ip routing-table
Route Flags: R - relay, D - download to fib
------------------------------------------------------------------
Routing Tables: Public
         Destinations : 14     Routes : 14
Destination/Mask    Proto   Pre  Cost  Flags NextHop      Interface
     10.0.0.0/8     Direct  0    0     D     10.0.0.1     GigabitEthernet0/0/0
     20.0.0.0/8     RIP     100  2     D     30.0.0.2     Serial1/0/0
     30.0.0.0/8     Direct  0    0     D     30.0.0.1     Serial1/0/0
     40.0.0.0/8     RIP     100  1     D     30.0.0.2     Serial1/0/0
     50.0.0.0/8     RIP     100  1     D     30.0.0.2     Serial1/0/0
     ……
```

结果显示，R1 仅配置直连的网段 10.0.0.0 和 30.0.0.0，经 RIP 作用后，R1 可以到达网段 20.0.0.0、40.0.0.0 和 50.0.0.0。

使用 display ip routing-table 命令查看 R2 的路由表。

```
<R2>display ip routing-table
Route Flags: R - relay, D - download to fib
```

计算机网络实践教程

```
------------------------------------------------------------------
Routing Tables: Public
        Destinations : 17      Routes : 17
Destination/Mask    Proto  Pre  Cost  Flags NextHop     Interface
     10.0.0.0/8     RIP    100   1      D   30.0.0.1    Serial1/0/0
     20.0.0.0/8     RIP    100   1      D   40.0.0.2    Serial1/0/1
     30.0.0.0/8     Direct 0     0      D   30.0.0.2    Serial1/0/0
     40.0.0.1/32    Direct 0     0      D   127.0.0.1   Serial1/0/1
     50.0.0.0/8     Direct 0     0      D   50.0.0.1    GigabitEthernet0/0/0
     ……
```

结果显示，R2 仅配置直连的网段 30.0.0.0、40.0.0.0 和 50.0.0.0，经 RIP 作用后，R1 可以到达网段 10.0.0.0 和 20.0.0.0。

使用 display ip routing-table 命令查看 R3 的路由表。

```
<R3>display ip routing-table
Route Flags: R - relay, D - download to fib
------------------------------------------------------------------
Routing Tables: Public
        Destinations : 14      Routes : 14
Destination/Mask    Proto  Pre  Cost  Flags NextHop     Interface
     10.0.0.0/8     RIP    100   2      D   40.0.0.1    Serial1/0/0
     20.0.0.0/8     Direct 0     0      D   20.0.0.1    GigabitEthernet0/0/0
     30.0.0.0/8     RIP    100   1      D   40.0.0.1    Serial1/0/0
     40.0.0.0/8     Direct 0     0      D   40.0.0.2    Serial1/0/0
     50.0.0.0/8     RIP    100   1      D   40.0.0.1    Serial1/0/0
……
```

结果显示，R3 仅配置直连的网段 20.0.0.0 和 40.0.0.0，经 RIP 作用后，R3 可以到达网段 10.0.0.0、30.0.0.0 和 50.0.0.0。

使用 ping 命令测试路由器 R1 与 R2 环回接口（非直连接口）间的连通性。

```
<R1>ping -c 1 40.0.0.1
 ping 40.0.0.1: 56 data bytes, press CTRL_C to break
   Reply from 40.0.0.1: bytes=56 Sequence=1 ttl=255 time=10 ms
 --- 40.0.0.1 ping statistics ---
   1 packet(s) transmitted
   1 packet(s) received
   0.00% packet loss
   round-trip min/avg/max = 10/10/10 ms
```

结果显示通信正常。同样地，可以测试 R1、R2 和 R3 其他环回接口间的连通性。

测试主机 PC1、PC2 和 PC3 间的连通性。在 PC1 上使用 ping 命令测试与 PC2 和 PC3 间的连通性。

```
PC>ping 20.0.0.2
ping 20.0.0.2: 32 data bytes, Press Ctrl_C to break
Request timeout!
From 20.0.0.2: bytes=32 seq=2 ttl=125 time=31 ms
From 20.0.0.2: bytes=32 seq=3 ttl=125 time=31 ms
From 20.0.0.2: bytes=32 seq=4 ttl=125 time=31 ms
From 20.0.0.2: bytes=32 seq=5 ttl=125 time=31 ms
--- 20.0.0.2 ping statistics ---
  5 packet(s) transmitted
  4 packet(s) received
  20.00% packet loss
  round-trip min/avg/max = 0/31/31 ms

PC>ping 50.0.0.2
ping 50.0.0.2: 32 data bytes, Press Ctrl_C to break
Request timeout!
From 50.0.0.2: bytes=32 seq=2 ttl=126 time=31 ms
From 50.0.0.2: bytes=32 seq=3 ttl=126 time=15 ms
From 50.0.0.2: bytes=32 seq=4 ttl=126 time=15 ms
From 50.0.0.2: bytes=32 seq=5 ttl=126 time=31 ms
--- 50.0.0.2 ping statistics ---
  5 packet(s) transmitted
  4 packet(s) received
  20.00% packet loss
  round-trip min/avg/max = 0/23/31 ms
```

结果显示通信正常。同样地，可以测试主机 PC1、PC2 和 PC3 相互之间的连通性。

使用 debuging 命令查看 RIP 协议定期更新情况，并开启 RIP 调试功能。请注意，debug 命令需要在用户视图下才能使用。使用 terminal debugging 和 terminal monitor 命令开启 debug 信息在屏幕上显示的功能，才能在计算机屏幕上看到路由器之间 RIP 协议交互的信息。

```
<R1>debugging rip 1
<R1>terminal debugging
Info: Current terminal debugging is on.
<R1>terminal monitor
<R1>
Jul 22 2022 09:11:36.125.2-08:00 R1 RIP/7/DBG:6:13456: RIP 1: Sending response on interface Serial1/0/0 from 30.0.0.1 to 255.255.255.255
<R1>
Jul 22 2022 09:11:36.125.3-08:00 R1 RIP/7/DBG:6:13476: Packet: Version 1, Cmd
response, Length 24
<R1>
```

```
    Jul 22 2022 09:11:36.125.4-08:00 R1 RIP/7/DBG:6:13527:Dest 10.0.0.0, Cost 1
    <R1>
    Jul 22 2022 09:11:42.95.1-08:00 R1 RIP/7/DBG:6:13414: RIP 1: Receiving v1
resp onse on Serial1/0/0 from 30.0.0.2 with 3 RTEs
    <R1>
    Jul 22 2022 09:11:42.95.2-08:00 R1 RIP/7/DBG:6:13465: RIP 1: Receive response
from 30.0.0.2 on Serial1/0/0
    <R1>
    Jul 22 2022 09:11:42.95.3-08:00 R1 RIP/7/DBG:6:13476: Packet:Version 1,Cmd
response, Length 64
    <R1>
    Jul 22 2022 09:11:42.95.4-08:00 R1 RIP/7/DBG:6:13527: Dest 20.0.0.0, Cost 2
    <R1>
    Jul 22 2022 09:11:42.95.5-08:00 R1 RIP/7/DBG:6:13527: Dest 40.0.0.0, Cost 1
    <R1>
    Jul 22 2022 09:11:42.95.6-08:00 R1 RIP/7/DBG:6:13527: Dest 50.0.0.0, Cost 1
    ……
    <R1>
    Jul 22 2022 09:12:10.105.2-08:00 R1 RIP/7/DBG:6:13465: RIP 1: Receive
response from 30.0.0.2 on Serial1/0/0
```

结果显示 R1 从连接 R2 的 Serial 1/0/0 接口周期性发送、接收 RIP v1 的 Response 更新报文，包括目的地、数据包大小以及 cost 值。

开启过多的调试功能将消耗路由器的大量资源，甚至可能导致宕机。可以使用 **undo debugging rip** 或 **undo debug all** 命令关闭 debug 调试功能。也可以使用带更多参数的命令查看某类型的调试信息，如 **debugging rip 1 event** 查看路由器发出和接收的定期更新事件。其他参数可以使用"？"获取帮助。

```
    <R1>debugging rip 1 event
    <R1>terminal debugging
    Info: Current terminal debugging is on.
    <R1>
    Jul 22 2022 11:57:35.885.1-08:00 R1 RIP/7/DBG:25:5719:RIP 1:Periodic timer
expired for interface Serial1/0/0 (30.0.0.1) and its added to periodic update
queue
    <R1>
    Jul 22 2022 11:57:35.885.2-08:00 R1 RIP/7/DBG: 25: 6048: RIP 1: Interface
Serial1/0/0 (30.0.0.1) is deleted from the periodic update queue
        <R1>undo debugging all
```

结果显示，路由器定期更新事件有开启。

4. 使用 RIPv2 搭建网络

基于前面的配置，现在只需要在 RIP 子视图模式下配置 RIP v2 即可。

```
[R1]rip
[R1-rip-1]version 2

[R2]rip
[R2-rip-1]version 2

[R2]rip
[R2-rip-1]version 2
```

配置完成后使用 display ip routing-table 命令查看各路由器路由表。使用 ping 命令检测 R1、R2 和 R3 之间直连链路的连通性，检测 PC1、PC2 和 PC3 之间的连通性。结果显示通信正常。使用 debugging 命令查看 RIPv2 定期更新情况。

```
<R1>debugging rip 2
<R1>terminal debugging
Info: Current terminal debugging is on.
<R1>terminal monitor
<R1>
Jul 22 2022 15:51:19.785.6-08:00 R1 RIP/7/DBG:25: 6048: RIP 1: Interface Serial
1/0/0 (30.0.0.1) is deleted from the periodic update queue
<R1>
Jul 22 2022 15:51:20.605.1-08:00 R1 RIP/7/DBG:6:13414: RIP 1:Receiving v2 res ponse on Serial1/0/0 from 30.0.0.2 with 3 RTEs
<R1>
Jul 22 2022 15:51:20.605.2-08:00 R1 RIP/7/DBG:6: 13465: RIP 1: Receive response from 30.0.0.2 on Serial1/0/0
<R1>
Jul 22 2022 15:51:20.605.3-08:00 R1 RIP/7/DBG:6:13476:Packet: Version 2, Cmd response, Length 64
<R1>
Jul 22 2022 15:51:20.605.4-08:00 R1 RIP/7/DBG:6:13546:Dest 20.0.0.0/8, Nextho p 0.0.0.0, Cost 2, Tag 0
<R1>
Jul 22 2022 15:51:20.605.5-08:00 R1 RIP/7/DBG:6:13546:Dest 40.0.0.0/8, Nextho p 0.0.0.0, Cost 1, Tag 0
<R1>
Jul 22 2022 15:51:20.605.6-08:00 R1 RIP/7/DBG:6:13546:Dest 50.0.0.0/8, Nextho p 0.0.0.0, Cost 1, Tag 0
……
<R1>
Jul 22 2022 15:51:20.975.2-08:00 R1 RIP/7/DBG:6: 13456: RIP 1: Sending response on interface Serial1/0/0 from 30.0.0.1 to 224.0.0.9
```

与 RIPv1 中使用 debugging 命令所查看的信息进行相比,可以明显区分出 RIPv1 和 RIPv2 的不同:

(1) RIPv2 的路由信息中携带了子网掩码。

(2) RIPv2 的路由信息中携带了下一跳地址,标识一个比通告路由器的地址更好的下一跳地址。换句话说,它指出的地址,其度量值(跳数)比在同一个网上的通告路由器更靠近目的地。如果这个字段设置为 0.0.0.0,说明通告的路由器的地址是最优的下一跳地址。

(3) RIPv2 默认采用组播方式发送报文,地址为 224.0.0.9。

5.3 OSPF 的动态路由配置

5.3.1 OSPF 单区域配置

一、原理概述

为了弥补距离矢量路由协议的不足,IETF 组织于 20 世纪 80 年代末开发了一种基于链路状态的内部网关协议 OSPF(open shortest path first,开放式最短路径优先)。

OSPF 协议具有收敛快、路由无环、扩展性好等优点,被迅速接受并广泛应用。链路状态算法路由协议互相通告的是链路状态信息,每台路由器都将自己的链路状态信息(包含接口的 IP 地址和子网掩码、网络类型、该链路的开销等)发送给其他路由器,并在网络中洪泛,当每台路由器收集到网络内所有链路状态信息后,就能拥有整个网络的拓扑情况,然后根据整个网络拓扑情况运行 SPF 算法,得出所有网段的最短路径。

OSPF 支持区域的划分,区域是从逻辑上将路由器划分为不同的组,每个组用区域号(area ID)来标识。一个网段只能属于一个区域,或者说每个运行 OSPF 的接口必须指明属于哪一个区域。区域 0 为骨干区域,骨干区域负责在非骨干区域之间发布区域间的路由信息。在一个 OSPF 区域中有且只有一个骨干区域。

二、实验目的

1. 理解 OSPF 单区域的应用场景
2. 掌握 OSPF 单区域的配置方法
3. 掌握查看 OSPF 邻居状态的方法

三、实验内容

本实验模拟企业网络场景。该公司有三大办公区,每个办公区放置了一台路由器,R1 放在办公区 A,A 区经理的 PC1 直接连接路由器 R1;路由器 R2 放在办公区 B,B 区经理的 PC2

直接连接到路由器 R2；R3 放在办公区 C，C 区经理的 PC3 直接连接到路由器 R3；3 台路由器都互相直连，为了能使整个公司网络互相通信，需要在所有路由器上部署路由协议。考虑到公司未来的发展（部门的增加和分公司的成立），为了适应不断扩展的网络需求，公司在所有路由器上部署 OSPF 协议，且现在所有路由器都属于骨干区域（单区域）。

四、网络拓扑

OSPF 单区域配置拓扑如图 5.11 所示。

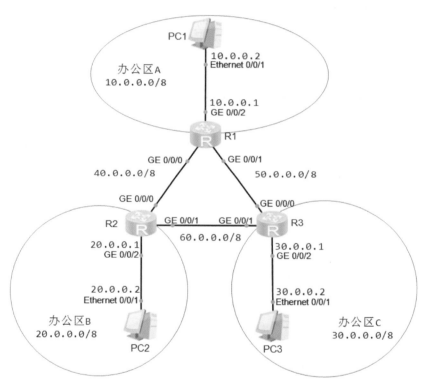

图 5.11 单区域 OSPF 配置拓扑

五、实验地址分配

本实验拓扑 IP 地址分配如表 5.3 所示。

表 5.3 IP 地址分配表

设备	接口	IP 地址	子网掩码	默认网关
PC1	Ethernet 0/0/1	10.0.0.2	255.0.0.0	10.0.0.1
PC2	Ethernet 0/0/1	20.0.0.2	255.0.0.0	20.0.0.1
PC3	Ethernet 0/0/1	30.0.0.2	255.0.0.0	30.0.0.1

续表

设备	接口	IP 地址	子网掩码	默认网关
R1（AR2220）	GE 0/0/0	40.0.0.1	255.0.0.0	N/A
	GE 0/0/1	50.0.0.1	255.0.0.0	N/A
	GE 0/0/2	10.0.0.1	255.0.0.0	N/A
R2（AR2220）	GE 0/0/0	40.0.0.2	255.0.0.0	N/A
	GE 0/0/1	60.0.0.1	255.0.0.0	N/A
	GE 0/0/2	20.0.0.1	255.0.0.0	N/A
R3（AR2220）	GE 0/0/0	50.0.0.2	255.0.0.0	N/A
	GE 0/0/1	60.0.0.2	255.0.0.0	N/A
	GE 0/0/2	30.0.0.1	255.0.0.0	N/A

六、实验步骤

1. PC 的基本配置

PC1 的基本配置如图 5.12 所示。

```
IPv4 配置
● 静态       ○ DHCP              □ 自动获取 DNS 服务器地址
IP 地址：    10 . 0 . 0 . 2      DNS1：  0 . 0 . 0 . 0
子网掩码：   255 . 0 . 0 . 0     DNS2：  0 . 0 . 0 . 0
网关：       10 . 0 . 0 . 1
```

图 5.12 PC1 的基本配置

PC2 的基本配置如图 5.13 所示。

```
IPv4 配置
● 静态       ○ DHCP              □ 自动获取 DNS 服务器地址
IP 地址：    20 . 0 . 0 . 2      DNS1：  0 . 0 . 0 . 0
子网掩码：   255 . 0 . 0 . 0     DNS2：  0 . 0 . 0 . 0
网关：       20 . 0 . 0 . 1
```

图 5.13 PC2 的基本配置

PC3 的基本配置如图 5.14 所示。

第 5 章 路由器配置

```
IPv4 配置
  ● 静态        ○ DHCP                      □ 自动获取 DNS 服务器地址
  IP 地址:    30 . 0 . 0 . 2       DNS1:    0 . 0 . 0 . 0
  子网掩码:   255 . 0 . 0 . 0      DNS2:    0 . 0 . 0 . 0
  网关:       30 . 0 . 0 . 1
```

图 5.14 PC3 的基本配置

2. 路由器的基本配置

根据实验地址分配表分别完成 R1、R2 和 R3 的基本配置。

1) R1 的配置

```
<Huawei>system-view
[Huawei]sysname R1
[R1]interface GigabitEthernet 0/0/0
[R1-GigabitEthernet0/0/0]ip address 40.0.0.1 255.0.0.0
[R1-GigabitEthernet0/0/0]interface GigabitEthernet 0/0/1
[R1-GigabitEthernet0/0/1]ip address 50.0.0.1 255.0.0.0
[R1-GigabitEthernet0/0/1]interface GigabitEthernet 0/0/2
[R1-GigabitEthernet0/0/2]ip address 10.0.0.1 255.0.0.0
```

2) R2 的配置

```
<Huawei>system-view
[Huawei]sysname R2
[R2]interface GigabitEthernet 0/0/0
[R2-GigabitEthernet0/0/0]ip address 40.0.0.2 255.0.0.0
[R2-GigabitEthernet0/0/0]interface GigabitEthernet 0/0/1
[R2-GigabitEthernet0/0/1]ip address 60.0.0.1 255.0.0.0
[R2-GigabitEthernet0/0/1]interface GigabitEthernet 0/0/2
[R2-GigabitEthernet0/0/2]ip address 20.0.0.1 255.0.0.0
```

3) R3 的配置

```
<Huawei>system-view
[Huawei]sysname R3
[R3]interface GigabitEthernet 0/0/0
[R3-GigabitEthernet0/0/0]ip address 50.0.0.2 255.0.0.0
[R3-GigabitEthernet0/0/0]interface GigabitEthernet 0/0/1
[R3-GigabitEthernet0/0/1]ip address 60.0.0.2 255.0.0.0
[R3-GigabitEthernet0/0/1]interface GigabitEthernet 0/0/2
[R3-GigabitEthernet0/0/2]ip address 30.0.0.1 255.0.0.0
```

计算机网络实践教程

在配置 OSPF 之前，使用 ping 命令检测直连链路的连通性。例如，检测 R1 到 R2 之间直连链路的连通性。

```
<R1>ping -c 1 40.0.0.2
  ping 40.0.0.2: 56  data bytes, press CTRL_C to break
    Reply from 40.0.0.2: bytes=56 Sequence=1 ttl=255 time=130 ms
  --- 40.0.0.2 ping statistics ---
    1 packet(s) transmitted
    1 packet(s) received
    0.00% packet loss
    round-trip min/avg/max = 130/130/130 ms
```

结果显示，各直连链路是连通的，即基本配置正确。

3. 部署单区域 OSPF 网络

在 R1 上使用 ospf 命令创建并运行 OSPF。

```
[R1]ospf 1
```

其中，1 代表的是进程号，如果没有写明进程号，则默认是 1。

使用 area 命令创建区域并进入 OSPF 区域视图，输入要创建的区域 ID。由于本实验为 OSPF 单区域配置，所以使用骨干区域，即区域 0 即可。

```
[R1-ospf-1]area 0
```

使用 network 命令来指定运行 OSPF 协议的接口和接口所属的区域。本实验中 R1 上的 3 个物理接口都需要指定。配置中需注意，尽量精确匹配所通告的网段。

```
[R1-ospf-1-area-0.0.0.0]network 10.0.0.0 0.255.255.255
[R1-ospf-1-area-0.0.0.0]network 40.0.0.0 0.255.255.255
[R1-ospf-1-area-0.0.0.0]network 50.0.0.0 0.255.255.255
```

配置完成后使用 display ospf interface 命令检查 OSPF 接口通告是否正确。

```
[R1]display ospf interface
OSPF Process 1 with Router ID 40.0.0.1
Interfaces
```

```
Area: 0.0.0.0            (MPLS TE not enabled)
IP Address   Type         State     Cost  Pri  DR        BDR
40.0.0.1     Broadcast    Waiting   1     1    0.0.0.0   0.0.0.0
50.0.0.1     Broadcast    Waiting   1     1    0.0.0.0   0.0.0.0
10.0.0.1     Broadcast    Waiting   1     1    0.0.0.0   0.0.0.0
```

结果显示，本地 OSPF 进程使用的 Router ID 是 40.0.0.1。在此进程下，有 3 个接口加入了 OSPF 进程。"Type"为以太网默认的广播网络类型；"State"为该接口当前的状态，显示为 DR 状态，即表示为这 3 个接口在它们所在的网段中都被选举为 DR。

接下来在 R2 和 R3 上做相应配置，配置方法和 R1 的相同。

```
[R2]ospf 1
[R2-ospf-1]area 0
[R2-ospf-1-area-0.0.0.0]network 20.0.0.0 0.255.255.255
[R2-ospf-1-area-0.0.0.0]network 40.0.0.0 0.255.255.255
[R2-ospf-1-area-0.0.0.0]network 60.0.0.0 0.255.255.255

[R3]ospf 1
[R3-ospf-1]area 0
[R3-ospf-1-area-0.0.0.0]network 30.0.0.0 0.255.255.255
[R3-ospf-1-area-0.0.0.0]network 50.0.0.0 0.255.255.255
[R3-ospf-1-area-0.0.0.0]network 60.0.0.0 0.255.255.255
```

4. 检查 OSPF 单区域的配置结果

```
<R1>display ospf peer
OSPF Process 1 with Router ID 40.0.0.1
Neighbors
 Area 0.0.0.0 interface 40.0.0.1(GigabitEthernet0/0/0)'s neighbors
 Router ID: 40.0.0.2       Address: 40.0.0.2
   State: Full  Mode:Nbr is Master  Priority: 1
   DR: 40.0.0.1 BDR: 40.0.0.2 MTU: 0
   Dead timer due in 35 sec
   Retrans timer interval: 5
   Neighbor is up for 00:06:36
   Authentication Sequence: [ 0 ]
Neighbors
 Area 0.0.0.0 interface 50.0.0.1(GigabitEthernet0/0/1)'s neighbors
 Router ID: 50.0.0.2       Address: 50.0.0.2
   State: Full  Mode:Nbr is Master  Priority: 1
   DR: 50.0.0.1 BDR: 50.0.0.2 MTU: 0
   Dead timer due in 30 sec
   Retrans timer interval: 5
   Neighbor is up for 00:03:48
```

```
Authentication Sequence: [ 0 ]
```

通过 display ospf peer 命令，可以查看很多内容。例如，通过 Router ID 可以查看邻居路由器标识；通过 Address 可以查看邻居 OSPF 接口 IP 地址；通过 State 可以查看目前与该路由器的 OSPF 邻居状态；通过 Priority 可以查看当前该邻居 OSPF 接口的 DR 优先级等。

使用 display ip routing-table protocol ospf 命令查看 R1 的 OSPF 路由表。

```
<R1>display ip routing-table protocol ospf
Route Flags: R - relay, D - download to fib
------------------------------------------------------------------
Public routing table : OSPF
        Destinations : 3        Routes : 4
OSPF routing table status : <Active>
        Destinations : 3        Routes : 4
Destination/Mask    Proto   Pre Cost  Flags NextHop    Interface
       20.0.0.0/8   OSPF    10   2      D  40.0.0.2   GigabitEthernet0/0/0
       30.0.0.0/8   OSPF    10   2      D  50.0.0.2   GigabitEthernet0/0/1
       60.0.0.0/8   OSPF    10   2      D  40.0.0.2   GigabitEthernet0/0/0
                    OSPF    10   2      D  50.0.0.2   GigabitEthernet0/0/1
OSPF routing table status : <Inactive>
        Destinations : 0        Routes : 0
```

结果显示，"Destination/Mask" 标识了目的网段的前缀及掩码，"Proto" 标识了此路由信息是通过 OSPF 协议获取的，"Pre" 标识了路由优先级，"Cost" 标识了开销值，"NextHop" 标识了下一跳地址，"Interface" 标识了此前缀的出接口。

此时 R1 的路由表中已经拥有了去往网络中所有其他网段的路由条目。用同样的方法查看 R2 和 R3 的 OSPF 邻居状态。

在 PC1 上使用 ping 命令测试与 PC2 和 PC3 之间的连通性。

```
PC>ping 20.0.0.2 -c 1
ping 20.0.0.2: 32 data bytes, Press Ctrl_C to break
From 20.0.0.2: bytes=32 seq=1 ttl=126 time=16 ms
--- 20.0.0.2 ping statistics ---
 1 packet(s) transmitted
 1 packet(s) received
 0.00% packet loss
 round-trip min/avg/max = 16/16/16 ms
```

5.3.2 OSPF 多区域配置

一、原理概述

在 OSPF 单区域中，每台路由器都需要收集其他所有路由器的链路状态信息，如果网络规模不断扩大，链路状态信息也会随之不断增多，这将使得单台路由器上链路状态数据库非常庞大，导致路由器负荷加重，也不便于维护管理。为了解决上述问题，OSPF 协议可以将整个自治系统划分为不同的区域。

链路状态信息只在区域内部洪泛，区域之间传递的只是路由条目而非链路状态信息，因此大大减少了路由器的负担。当一台路由器的接口属于不同区域时称它为区域边界路由器（area border router，ABR），负责传递区域间路由信息。区域间的路由信息传递类似距离矢量算法，为了防止区域间产生环路，所有非骨干区域之间的路由信息必须经过骨干区域，也就是说非骨干区域必须和骨干区域相连，且非骨干区域之间不能直接进行路由信息交换。

二、实验目的

1. 理解 OSPF 多区域的应用场景
2. 掌握 OSPF 多区域的配置方法
3. 理解 OSPF 区域边界路由器（ABR）的工作特点

三、实验内容

本实验模拟企业网络场景。路由器 R1、R2、R3、R4 为企业总部核心区域设备，属于区域 0，R5 属于新增分支机构 A 的网关设备，R6 属于新增分支机构 B 的网关设备。PC1 和 PC2 分别属于分支机构 A 和 B，PC3 和 PC4 属于总部管理员登录设备，用于网络管理。

在该网络中，如果设计方案采用单区域配置，则会导致单一区域 LSA 数目过大，导致路由器开销过高，SPF 算法运算过于频繁。因此，网络管理员选择配置多区域方案进行网络配置，将两个新分支运行在不同的 OSPF 区域中，其中 R5 属于区域 1，R6 属于区域 2。

四、实验拓扑

OSPF 多区域配置拓扑如图 5.15 所示。

注：R3 和 R4 分别增加 1 个接口卡 "2 端口-FE WAN 接口卡"，用于连接 PC3 和 PC4。

五、实验地址分配

本实验地址分配如表 5.4 所示。

计算机网络实践教程

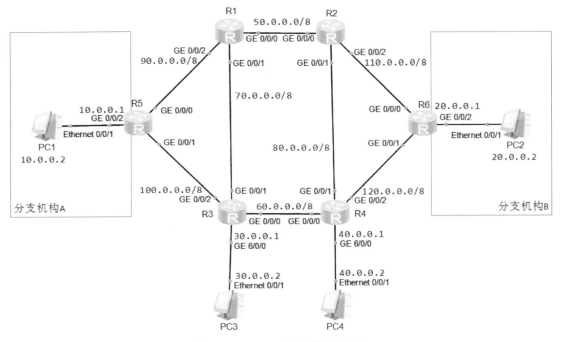

图 5.15 OSPF 多区域配置拓扑

表 5.4 实验 IP 地址分配表

设备	接口	IP 地址	子网掩码	默认网关
PC1	Ethernet 0/0/1	10.0.0.2	255.0.0.0	10.0.0.1
PC2	Ethernet 0/0/1	20.0.0.2	255.0.0.0	20.0.0.1
PC3	Ethernet 0/0/1	30.0.0.2	255.0.0.0	30.0.0.1
PC4	Ethernet 0/0/1	40.0.0.2	255.0.0.0	40.0.0.1
R1（AR2240）	GE 0/0/0	50.0.0.1	255.0.0.0	N/A
	GE 0/0/1	70.0.0.1	255.0.0.0	N/A
	GE 0/0/2	90.0.0.2	255.0.0.0	N/A
R2（AR2240）	GE 0/0/0	50.0.0.2	255.0.0.0	N/A
	GE 0/0/1	80.0.0.1	255.0.0.0	N/A
	GE 0/0/2	110.0.0.1	255.0.0.0	N/A
R3（AR2240）	GE 0/0/0	60.0.0.1	255.0.0.0	N/A
	GE 0/0/1	70.0.0.2	255.0.0.0	N/A
	GE 0/0/2	100.0.0.2	255.0.0.0	N/A
	Ethernet 1/0/0	30.0.0.1	255.0.0.0	N/A

续表

设备	接口	IP 地址	子网掩码	默认网关
R4（AR2240）	GE 0/0/0	60.0.0.2	255.0.0.0	N/A
	GE 0/0/1	80.0.0.2	255.0.0.0	N/A
	GE 0/0/2	120.0.0.1	255.0.0.0	N/A
	Ethernet 1/0/0	40.0.0.1	255.0.0.0	N/A
R5（AR2240）	GE 0/0/0	90.0.0.1	255.0.0.0	N/A
	GE 0/0/1	100.0.0.1	255.0.0.0	N/A
	GE 0/0/2	10.0.0.1	255.0.0.0	N/A
R6（AR2240）	GE 0/0/0	110.0.0.2	255.0.0.0	N/A
	GE 0/0/1	120.0.0.2	255.0.0.0	N/A
	GE 0/0/2	20.0.0.1	255.0.0.0	N/A

六、实验步骤

1. PC 的基本配置

PC1 的基本配置如图 5.16 所示。

图 5.16　PC1 的基本配置

PC2 的基本配置如图 5.17 所示。

图 5.17　PC2 的基本配置

PC3 的基本配置如图 5.18 所示。

```
IPv4 配置
● 静态        ○ DHCP              □ 自动获取 DNS 服务器地址
IP 地址:       30 . 0 . 0 . 2     DNS1:      0 . 0 . 0 . 0
子网掩码:     255 . 0 . 0 . 0     DNS2:      0 . 0 . 0 . 0
网关:          30 . 0 . 0 . 1
```

图 5.18　PC3 的基本配置

PC4 的基本配置如图 5.19 所示。

```
IPv4 配置
● 静态        ○ DHCP              □ 自动获取 DNS 服务器地址
IP 地址:       40 . 0 . 0 . 2     DNS1:      0 . 0 . 0 . 0
子网掩码:     255 . 0 . 0 . 0     DNS2:      0 . 0 . 0 . 0
网关:          40 . 0 . 0 . 1
```

图 5.19　PC4 的基本配置

2. 路由器的基本配置

根据实验地址分配表分别完成 R1、R2、R3、R4、R5 和 R6 的基本配置。

1) R1 的配置

```
<Huawei>system-view
[Huawei]sysname R1
[R1]interface GigabitEthernet 0/0/0
[R1-GigabitEthernet0/0/0]ip address 50.0.0.1 255.0.0.0
[R1-GigabitEthernet0/0/0]interface GigabitEthernet 0/0/1
[R1-GigabitEthernet0/0/1]ip address 70.0.0.1 255.0.0.0
[R1-GigabitEthernet0/0/1]interface GigabitEthernet 0/0/2
[R1-GigabitEthernet0/0/2]ip address 90.0.0.2 255.0.0.0
```

2) R2 的配置

```
<Huawei>system-view
[Huawei]sysname R2
[R2]interface GigabitEthernet 0/0/0
[R2-GigabitEthernet0/0/0]ip address 50.0.0.2 255.0.0.0
[R2-GigabitEthernet0/0/0]interface GigabitEthernet 0/0/1
[R2-GigabitEthernet0/0/1]ip address 80.0.0.1 255.0.0.0
[R2-GigabitEthernet0/0/1]interface GigabitEthernet 0/0/2
[R2-GigabitEthernet0/0/2]ip address 110.0.0.1 255.0.0.0
```

3) R3 的配置

```
<Huawei>system-view
[Huawei]sysname R3
[R3]interface GigabitEthernet 0/0/0
[R3-GigabitEthernet0/0/0]ip address 60.0.0.1 255.0.0.0
[R3-GigabitEthernet0/0/0]interface GigabitEthernet 0/0/1
[R3-GigabitEthernet0/0/1]ip address 70.0.0.2 255.0.0.0
[R3-GigabitEthernet0/0/1]interface GigabitEthernet 0/0/2
[R3-GigabitEthernet0/0/2]ip address 100.0.0.2 255.0.0.0
[R3-GigabitEthernet0/0/2]interface Ethernet 1/0/0
[R3--Ethernet1/0/0]ip address 30.0.0.1 255.0.0.0
```

4) R4 的配置

```
<Huawei>system-view
[Huawei]sysname R4
[R4]interface GigabitEthernet 0/0/0
[R4-GigabitEthernet0/0/0]ip address 60.0.0.2 255.0.0.0
[R4-GigabitEthernet0/0/0]interface GigabitEthernet 0/0/1
[R4-GigabitEthernet0/0/1]ip address 80.0.0.2 255.0.0.0
[R4-GigabitEthernet0/0/1]interface GigabitEthernet 0/0/2
[R4-GigabitEthernet0/0/2]ip address 120.0.0.1 255.0.0.0
[R4-GigabitEthernet0/0/2]interface Ethernet 1/0/0
[R4--Ethernet1/0/0]ip address 40.0.0.1 255.0.0.0
```

5) R5 的配置

```
<Huawei>system-view
[Huawei]sysname R5
[R5]interface GigabitEthernet 0/0/0
[R5-GigabitEthernet0/0/0]ip address 90.0.0.1 255.0.0.0
[R5-GigabitEthernet0/0/0]interface GigabitEthernet 0/0/1
[R5-GigabitEthernet0/0/1]ip address 100.0.0.1 255.0.0.0
[R5-GigabitEthernet0/0/1]interface GigabitEthernet 0/0/2
[R5-GigabitEthernet0/0/2]ip address 10.0.0.1 255.0.0.0
```

6) R6 的配置

```
<Huawei>system-view
[Huawei]sysname R6
```

```
[R6]interface GigabitEthernet 0/0/0
[R6-GigabitEthernet0/0/0]ip address 110.0.0.2 255.0.0.0
[R6-GigabitEthernet0/0/0]interface GigabitEthernet 0/0/1
[R6-GigabitEthernet0/0/1]ip address 120.0.0.2 255.0.0.0
[R6-GigabitEthernet0/0/1]interface GigabitEthernet 0/0/2
[R6-GigabitEthernet0/0/2]ip address 20.0.0.1 255.0.0.0
```

在配置 OSPF 之前，使用 ping 命令检测直连链路的连通性。例如，检测 R1 到 R2 间直连链路的连通性。

```
<R1>ping -c 1 50.0.0.2
 ping 50.0.0.2: 56  data bytes, press CTRL_C to break
   Reply from 50.0.0.2: bytes=56 Sequence=1 ttl=255 time=20 ms
 --- 50.0.0.2 ping statistics ---
   1 packet(s) transmitted
   1 packet(s) received
   0.00% packet loss
   round-trip min/avg/max = 20/20/20 ms
```

测试路由器 R1 与 R2 之间正常通信，同样可以测试其余直连网段的连通性。

3. 配置骨干区域路由器

在公司总部路由器 R1、R2、R3 和 R4 上创建 OSPF 进程，并在骨干区域 0 视图下通告总部各网段。

在各路由器上，使用 **osfp 1** 命令创建并运行 OSPF，进程号为 1，使用 **area 0** 命令将路由器在骨干区域 0 视图下进行通告，通告与本路由器直连的网段。

```
[R1]ospf 1
[R1-ospf-1]area 0
[R1-ospf-1-area-0.0.0.0]network 50.0.0.0 0.255.255.255
[R1-ospf-1-area-0.0.0.0]network 70.0.0.0 0.255.255.255

[R2]ospf 1
[R2-ospf-1]area 0
[R2-ospf-1-area-0.0.0.0]network 50.0.0.0 0.255.255.255
[R2-ospf-1-area-0.0.0.0]network 80.0.0.0 0.255.255.255

[R3]ospf 1
[R3-ospf-1]area 0
[R3-ospf-1-area-0.0.0.0]network 70.0.0.0 0.255.255.255
[R3-ospf-1-area-0.0.0.0]network 60.0.0.0 0.255.255.255
[R3-ospf-1-area-0.0.0.0]network 30.0.0.0 0.255.255.255

[R4]ospf 1
```

```
[R4-ospf-1]area 0
[R4-ospf-1-area-0.0.0.0]network 80.0.0.0 0.255.255.255
[R4-ospf-1-area-0.0.0.0]network 60.0.0.0 0.255.255.255
[R4-ospf-1-area-0.0.0.0]network 40.0.0.0 0.255.255.255
```

配置完成后,测试总部内两台 PC 之间的连通性。

```
PC>ping 40.0.0.2 -c 1
ping 40.0.0.2: 32 data bytes, Press Ctrl_C to break
From 40.0.0.2: bytes=32 seq=1 ttl=126 time=15 ms
--- 40.0.0.2 ping statistics ---
 1 packet(s) transmitted
 1 packet(s) received
 0.00% packet loss
 round-trip min/avg/max = 15/15/15 ms
```

结果显示 PC3 和 PC4 之间正常通信。

4. 配置非骨干区域路由器

在分支机构 A 的 R5 上创建 OSPF 进程,创建并进入区域 1,并通告分支机构 A 的相应网段。

```
[R5]ospf 1
[R5-ospf-1]area 1
[R5-ospf-1-area-0.0.0.1]network 90.0.0.0 0.255.255.255
[R5-ospf-1-area-0.0.0.1]network 100.0.0.0 0.255.255.255
[R5-ospf-1-area-0.0.0.1]network 10.0.0.0 0.255.255.255
```

在 R1 和 R3 上也创建并进入区域 1 视图,将与 R5 相连的接口进行通告。

```
[R1]ospf 1
[R1-ospf-1]area 1
[R1-ospf-1-area-0.0.0.1]network 90.0.0.0 0.255.255.255

[R3]ospf 1
[R3-ospf-1]area 1
[R3-ospf-1-area-0.0.0.1]network 100.0.0.0 0.255.255.255
```

配置完成后,在路由器 R5 上使用 display ospf peer 命令查看 OSPF 邻居状态。

```
[R5]display ospf peer
OSPF Process 1 with Router ID 90.0.0.1
Neighbors
Area 0.0.0.1 interface 90.0.0.1(GigabitEthernet0/0/0)'s neighbors
Router ID: 50.0.0.1          Address: 90.0.0.2
  State: Full  Mode:Nbr is Slave  Priority: 1
  DR: 90.0.0.1  BDR: 90.0.0.2  MTU: 0
  ……
Neighbors
Area 0.0.0.1 interface 100.0.0.1(GigabitEthernet0/0/1)'s neighbors
Router ID: 60.0.0.1          Address: 100.0.0.2
  State: Full  Mode:Nbr is Slave  Priority: 1
  DR: 100.0.0.1  BDR: 100.0.0.2  MTU: 0
  ……
```

结果显示，现在 R5 与 R1 和 R3 的 OSPF 邻居关系建立正常，都为 Full 状态。

使用 display ip routing-table protocol ospf 命令查看 R5 路由表中 OSPF 路由条目。

```
[R5]display ip routing-table protocol ospf
Route Flags: R - relay, D - download to fib
------------------------------------------------------------------
Public routing table : OSPF
         Destinations : 6       Routes : 8
OSPF routing table status : <Active>
         Destinations : 6       Routes : 8
Destination/Mask    Proto  Pre  Cost  Flags  NextHop      Interface
      30.0.0.0/8    OSPF   10   2     D      100.0.0.2    GigabitEthernet0/0/1
      40.0.0.0/8    OSPF   10   3     D      100.0.0.2    GigabitEthernet0/0/1
      50.0.0.0/8    OSPF   10   2     D      90.0.0.2     GigabitEthernet0/0/0
      60.0.0.0/8    OSPF   10   2     D      100.0.0.2    GigabitEthernet0/0/1
      70.0.0.0/8    OSPF   10   2     D      90.0.0.2     GigabitEthernet0/0/0
                    OSPF   10   2     D      100.0.0.2    GigabitEthernet0/0/1
      80.0.0.0/8    OSPF   10   3     D      90.0.0.2     GigabitEthernet0/0/0
                    OSPF   10   3     D      100.0.0.2    GigabitEthernet0/0/1
```

结果显示，除 OSPF 区域 2 内的路由外，相关 OSPF 路由条目都已经获得。在拓扑中，R1 和 R3 这两台连接不同区域的路由器称为 ABR，即区域边界路由器，这类路由器设备可以同时属于两个以上的区域，但其中至少一个端口必须在骨干区域内。ABR 是用来连接骨干区域和非骨干区域的，其与骨干区域之间既可以是物理连接，也可以是逻辑上的连接。

使用 display ospf lsdb 命令查看 R5 的 OSPF 链路状态数据库信息。

```
[R5]display ospf lsdb
 OSPF Process 1 with Router ID 90.0.0.1
```

```
Link State Database
Area: 0.0.0.1
Type          LinkState ID      AdvRouter         Age    Len    Sequence      Metric
Router        50.0.0.1          50.0.0.1          414    36     80000004      1
Router        60.0.0.1          60.0.0.1          360    36     80000004      1
Router        90.0.0.1          90.0.0.1          350    60     8000000D      1
Network       90.0.0.1          90.0.0.1          412    32     80000003      0
Network       100.0.0.1         90.0.0.1          350    32     80000003      0
Sum-Net       50.0.0.0          50.0.0.1          423    28     80000002      1
Sum-Net       50.0.0.0          60.0.0.1          360    28     80000002      2
Sum-Net       70.0.0.0          50.0.0.1          423    28     80000002      1
Sum-Net       70.0.0.0          60.0.0.1          360    28     80000002      1
Sum-Net       40.0.0.0          50.0.0.1          423    28     80000002      3
Sum-Net       40.0.0.0          60.0.0.1          360    28     80000002      2
Sum-Net       60.0.0.0          50.0.0.1          423    28     80000002      2
Sum-Net       60.0.0.0          60.0.0.1          360    28     80000002      1
Sum-Net       30.0.0.0          50.0.0.1          423    28     80000002      2
Sum-Net       30.0.0.0          60.0.0.1          360    28     80000002      1
Sum-Net       80.0.0.0          50.0.0.1          423    28     80000002      2
Sum-Net       80.0.0.0          60.0.0.1          360    28     80000002      2
```

结果显示，关于其他区域的路由条目都是通过 "Sum-Net" 这类 LSA 获得，而这类 LSA 是不参与本区域的 SPF 算法运算的。

对公司另一分支机构 B 的路由器 R6，与相应 ABR 设备 R2 和 R4 也做同样的配置。

```
[R6]ospf 1
[R6-ospf-1]area 2
[R6-ospf-1-area-0.0.0.2]network 110.0.0.0 0.255.255.255
[R6-ospf-1-area-0.0.0.2]network 120.0.0.0 0.255.255.255
[R6-ospf-1-area-0.0.0.2]network 20.0.0.0 0.255.255.255

[R2]ospf 1
[R2-ospf-1]area 2
[R2-ospf-1-area-0.0.0.2]network 110.0.0.0 0.255.255.255

[R4]ospf 1
[R4-ospf-1]area 2
[R4-ospf-1-area-0.0.0.2]network 120.0.0.0 0.255.255.255
```

配置完成，在路由器 R6 上使用 display ip routing-table protocol ospf 命令查看 R6 的 OSPF 路由条目。

```
<R6>display ip routing-table protocol ospf
Route Flags: R - relay, D - download to fib
------------------------------------------------------------------------
```

```
Public routing table : OSPF
 Destinations : 9       Routes : 12
OSPF routing table status : <Active>
 Destinations : 9       Routes : 12
Destination/Mask  Proto  Pre  Cost  Flags  NextHop       Interface
    10.0.0.0/8    OSPF   10   4     D      110.0.0.1     GigabitEthernet0/0/0
                  OSPF   10   4     D      120.0.0.1     GigabitEthernet0/0/1
    30.0.0.0/8    OSPF   10   3     D      120.0.0.1     GigabitEthernet0/0/1
    40.0.0.0/8    OSPF   10   2     D      120.0.0.1     GigabitEthernet0/0/1
    50.0.0.0/8    OSPF   10   2     D      110.0.0.1     GigabitEthernet0/0/0
    60.0.0.0/8    OSPF   10   2     D      120.0.0.1     GigabitEthernet0/0/1
    70.0.0.0/8    OSPF   10   3     D      110.0.0.1     GigabitEthernet0/0/0
                  OSPF   10   3     D      120.0.0.1     GigabitEthernet0/0/1
    80.0.0.0/8    OSPF   10   2     D      110.0.0.1     GigabitEthernet0/0/0
                  OSPF   10   2     D      120.0.0.1     GigabitEthernet0/0/1
    90.0.0.0/8    OSPF   10   3     D      110.0.0.1     GigabitEthernet0/0/0
   100.0.0.0/8    OSPF   10   3     D      120.0.0.1     GigabitEthernet0/0/1
OSPF routing table status : <Inactive>
     Destinations : 0       Routes : 0
```

观察到可以正常接收到所有 OSPF 路由信息。测试分支机构 A 和 B 的主机之间的连通性。在主机 PC1 上使用 ping 命令测试与 PC2 之间的连通性。

```
PC>ping 20.0.0.2 -c 1
ping 20.0.0.2: 32 data bytes, Press Ctrl_C to break
From 20.0.0.2: bytes=32 seq=1 ttl=124 time=31 ms
--- 20.0.0.2 ping statistics ---
 1 packet(s) transmitted
 1 packet(s) received
 0.00% packet loss
 round-trip min/avg/max = 31/31/31 ms
```

结果显示，PC1 与 PC2 之间通信正常。

5.4　BGP 邻居配置

一、原理概述

路由协议通常分为内部网关协议 IGP 和外部网关协议 EGP 两大类。IGP 用于自治系统 AS

内部，EGP 用于 AS 之间。目前常见的 IGP 包括 RIP、OSPF 等，EGP 包括 BGP（border gateway protocol）等。

BGP 提供了丰富的路由属性（attribute），通过对这些属性的操作和控制，BGP 能够非常容易地实现丰富而灵活的路由策略。BGP 还具有良好的扩展性，支持 Multicast、VPN、IPv6 等多种属性。

BGP 的邻居关系分为 IBGP 和 EBGP 两种。当两台 BGP 路由器位于同一 AS 时，它们的邻居关系为 IBGP 关系；当两台 BGP 路由器位于不同 AS 时，它们的邻居关系为 EBGP 关系。BGP 没有自动建立邻居关系的能力，邻居关系必须通过手动配置来建立。

二、实验目的

1. 掌握 IBGP 与 EBGP 邻居的概念
2. 配置 IBGP 与 EBGP 邻居的关系

三、实验内容

实验模拟通过 ISP 之间的网络场景。R1 与 R2 属于同一个 ISP 网络，AS 编号为 100，R1 和 R2 之间的邻居关系为 IBGP 邻居关系。R3 属于另一个 ISP，AS 编号为 200，R3 与 R2 之间的邻居关系为 EBGP 邻居关系。本实验的路由器分别采用物理接口和 Loopback 接口来进行 IBGP 和 EBGP 邻居关系的建立。

四、实验拓扑

BGP 邻居配置拓扑如图 5.20 所示。

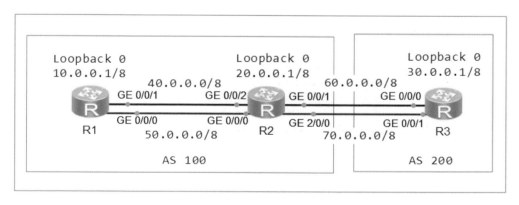

图 5.20　BGP 邻居配置拓扑

五、实验地址分配

本实验地址分配如表 5.5 所示。

表 5.5 实验地址分配表

设备	接口	IP 地址	子网掩码	默认网关
R1（AR2220）	GE 0/0/0	50.0.0.1	255.0.0.0	N/A
	GE 0/0/1	40.0.0.1	255.0.0.0	N/A
	Loopback 0	10.0.0.1	255.0.0.0	N/A
R2（AR2220）	GE 0/0/0	50.0.0.2	255.0.0.0	N/A
	GE 0/0/1	60.0.0.1	255.0.0.0	N/A
	GE 0/0/2	40.0.0.2	255.0.0.0	N/A
	GE 2/0/0	70.0.0.1	255.0.0.0	N/A
	Loopback 0	20.0.0.1	255.0.0.0	N/A
R3（AR2220）	GE 0/0/0	60.0.0.2	255.0.0.0	N/A
	GE 0/0/1	70.0.0.2	255.0.0.0	N/A
	Loopback 0	30.0.0.1	255.0.0.0	N/A

六、实验步骤

1. 路由器的基本配置

根据实验地址分配表进行相应的基本配置。

1) R1 的配置

```
<Huawei>system-view
[Huawei]sysname R1
[R1]int loopback 0
[R1-LoopBack0]ip add 10.0.0.1 255.0.0.0
[R1-LoopBack0]int GigabitEthernet 0/0/0
[R1-GigabitEthernet0/0/0]ip address 50.0.0.1 255.0.0.0
[R1-GigabitEthernet0/0/0]int GigabitEthernet 0/0/1
[R1-GigabitEthernet0/0/1]ip address 40.0.0.1 255.0.0.0
```

2) R2 的配置

R2 增加 1 个接口卡 "4 端口-GE 电口 WAN 接口卡"。

```
<Huawei>system-view
[Huawei]sysname R2
[R2]int loopback 0
[R2-LoopBack0]ip add 20.0.0.1 255.0.0.0
[R2-LoopBack0]int GigabitEthernet 0/0/0
```

```
[R2-GigabitEthernet0/0/0]ip address 50.0.0.2 255.0.0.0
[R2-GigabitEthernet0/0/0]int GigabitEthernet 0/0/1
[R2-GigabitEthernet0/0/1]ip address 60.0.0.1 255.0.0.0
[R2-GigabitEthernet0/0/1]int GigabitEthernet 0/0/2
[R2-GigabitEthernet0/0/2]ip address 40.0.0.2 255.0.0.0
[R2-GigabitEthernet0/0/2]int GigabitEthernet 2/0/0
[R2-GigabitEthernet2/0/0]ip address 70.0.0.1 255.0.0.0
```

3) R3 的配置

```
<Huawei>system-view
[Huawei]sysname R3
[R3]int loopback 0
[R3-LoopBack0]ip add 30.0.0.1 255.0.0.0
[R3-LoopBack0]int GigabitEthernet 0/0/0
[R3-GigabitEthernet0/0/0]ip address 60.0.0.2 255.0.0.0
[R3-GigabitEthernet0/0/0]int GigabitEthernet 0/0/1
[R3-GigabitEthernet0/0/1]ip address 70.0.0.2 255.0.0.0
```

使用 ping 命令检测 R1 与 R2 之间的连通性。用同样的方法测试其他直连链路的连通性。

```
PC>ping 40.0.0.2 -c 1
ping 20.0.0.2: 32 data bytes, Press Ctrl_C to break
From 20.0.0.2: bytes=32 seq=1 ttl=124 time=31 ms
--- 20.0.0.2 ping statistics ---
 1 packet(s) transmitted
 1 packet(s) received
 0.00% packet loss
 round-trip min/avg/max = 31/31/31 ms
```

2. 配置 IBGP 邻居

接下来将 R1 和 R2 上使用直连物理接口来配置 IBGP 邻居关系。为了实现链路冗余，R1 和 R2 之间部署了两条链路，当其中一条物理链路出现故障时，另一条物理链路可以提供连通性。

```
[R1]bgp 100
[R1-bgp]router-id 10.0.0.1
[R1-bgp]peer 40.0.0.2 as-number 100
[R1-bgp]peer 50.0.0.2 as-number 100

[R2]bgp 100
[R2-bgp]router-id 20.0.0.1
[R2-bgp]peer 40.0.0.1 as-number 100
```

```
[R2-bgp]peer 50.0.0.1 as-number 100
```

完成配置后,在 R2 上使用 display bgp peer 命令查看 BGP 邻居关系。

```
[R2]display bgp peer
 BGP local router ID : 20.0.0.1
 Local AS number : 100
 Total number of peers : 2   Peers in established state : 2
   Peer       V  AS    MsgRcvd  MsgSent  OutQ  Up/Down   State       PrefRcv
   40.0.0.1  4  100     4        4        0    00:02:16  Established   0
   50.0.0.1  4  100     4        4        0    00:02:07  Established   0
```

结果显示,R2 现在有两个 BGP 邻居,分别使用了 R1 的 GE 0/0/1 和 GE 0/0/0 接口地址来表示,AS 编号为 100,与 R2 自己的 AS 编号相同,因此 R2 和 R1 为 IBGP 邻居。当前邻居状态为 Established,表示邻居关系已完成建立。

在 R1 上将 Loopback 0 接口地址通告到 BGP 路由表。

```
[R1]bgp 100
[R1-bgp]network 10.0.0.1 8
```

上述配置完成后,在 R2 上使用 display bgp routing-table 命令查看 BGP 路由表。

```
<R2>display bgp routing-table
 BGP Local router ID is 20.0.0.1
 Status codes: * - valid, > - best, d - damped,
               h - history, i - internal, s - suppressed, S - Stale
               Origin : i - IGP, e - EGP, ? - incomplete
 Total Number of Routes: 2
     Network        NextHop      MED    LocPrf    PrefVal  Path/Ogn
 *>i 10.0.0.0       40.0.0.1      0      100        0        i
 *  i                50.0.0.1      0      100        0        i
```

结果显示,R2 的 BGP 路由表中包含了两条去往 10.0.0.0/8 的路由,下一跳分别为 40.0.0.1 和 50.0.0.1,这是因为 R1 与 R2 之间建立了两个 IBGP 邻居关系,BGP 路由实现了冗余。

BGP 运行在 TCP 之上,如果能让 R1 的 Loopback 0 接口与 R2 的 Loopback 0 接口建立起 TCP 会话,并使用 Loopback 0 接口的 IP 地址来建立 BGP 邻居关系,则可以让 R1 和 R2 只维护一个 BGP 邻居关系即可。当 R1 与 R2 之间的一条链路出现故障时,TCP 可以通过另外一条物理链路继续维持会话关系,这种方法在网络的稳定性方面和网络资源的节省上比直接使用物理接口来建立 BGP 邻居关系更具优势。

为了能使 R1 的 Loopback 0 接口与 R2 的 Loopback 0 接口建立起 TCP 会话，需要在 R1 和 R2 上配置到达对方 Loopback 0 接口的静态路由。

```
[R1]ip route-static 20.0.0.0 8 40.0.0.2
[R1]ip route-static 20.0.0.0 8 50.0.0.2

[R2]ip route-static 10.0.0.0 8 40.0.0.1
[R2]ip route-static 10.0.0.0 8 50.0.0.1
```

删除之前采用物理接口配置 IBGP 邻居的命令，并使用 Loopback 0 接口重新建立 IBGP 邻居关系。

```
[R1]bgp 100
[R1-bgp]undo peer 40.0.0.2
[R1-bgp]undo peer 50.0.0.2
[R1-bgp]peer 20.0.0.1 as-number 100

[R2]bgp 100
[R2-bgp]undo peer 40.0.0.1
[R2-bgp]undo peer 50.0.0.1
[R2-bgp]peer 10.0.0.1 as-number 100
```

配置完成后，在 R1 上使用 display bgp peer 命令查看 BGP 邻居关系。

```
[R1]display bgp peer
 BGP local router ID : 10.0.0.1
 Local AS number : 100
 Total number of peers : 1    Peers in established state : 0
  Peer        V    AS   MsgRcvd   MsgSent   OutQ   Up/Down   State    PrefRcv
  20.0.0.1    4    100  0         0         0      00:01:47  Active   0
```

结果显示，R1 和 R2 的邻居关系停留在 Active 状态，而非 Established，这说明 R1 与 R2 尚未正常建立起 IBGP 邻居关系。

在配置 BGP 邻居时所使用的 IP 地址，应该互为 BGP 报文的源 IP 地址和目的 IP 地址。默认情况下，BGP 会使用去往邻居路由器的出接口的 IP 地址作为 BGP 报文的源地址。在上面的配置中，R2 向 R1 发送 BGP 报文的源地址和 R1 上指定的邻居地址 20.0.0.1 不一致，从而导致 R1 无法和 R2 正常建立 BGP 邻居关系。解决此问题的方法是通过命令来强制指定路由器发送 BGP 报文时所使用的源 IP 地址。

在 R1 上使用 **peer 20.0.0.1 connect-interface LoopBack 0** 命令指定 R1 使用自己的 loopback 0 接口地址作为发送 BGP 报文时的源 IP 地址；R2 上也需要使用类似的命令。

```
[R1]bgp 100
[R1-bgp]peer 20.0.0.1 connect-interface loopback 0

[R2]bgp 100
[R2-bgp]peer 10.0.0.1 connect-interface loopback 0
```

配置完成后，在 R2 上使用 display bgp peer 命令查看 BGP 邻居关系。

```
[R2]display bgp peer
BGP local router ID : 20.0.0.1
Local AS number : 100
Total number of peers : 1      Peers in established state : 1
 Peer        V  AS   MsgRcvd  MsgSent  OutQ  Up/Down   State       PrefRcv
 10.0.0.1    4  100  6        5        0     00:03:01  Established 1
```

结果显示，R2 现在只与 10.0.0.1 有一个 IBGP 邻居关系，状态为 Established。
R2 上使用 display bgp routing-table 命令查看 BGP 路由表。

```
[R2]display bgp routing-table
BGP Local router ID is 20.0.0.1
Status codes: * - valid, > - best, d - damped,
              h - history, i - internal, s - suppressed, S - Stale
              Origin : i - IGP, e - EGP, ? - incomplete
Total Number of Routes: 1
   Network        NextHop       MED      LocPrf    PrefVal Path/Ogn
i  10.0.0.0       10.0.0.1      0        100       0       i
```

结果显示，R2 的 BGP 路由表中只有一条去往 10.0.0.0/8 的路由，下一跳为 10.0.0.1。
再次查看 R2 的路由表。

```
[R2]display ip routing-table
Route Flags: R - relay, D - download to fib
------------------------------------------------------------------------
Routing Tables: Public
         Destinations : 20     Routes : 21
Destination/Mask   Proto   Pre  Cost  Flags  NextHop    Interface
    10.0.0.0/8    Static  60   0     RD     40.0.0.1   GigabitEthernet0/0/2
                  Static  60   0     RD     50.0.0.1   GigabitEthernet0/0/0
    20.0.0.0/8    Direct  0    0     D      20.0.0.1   LoopBack0
    ……
```

结果显示，R2 去往 10.0.0.1 的路由有两条，下一跳分别是 40.0.0.1 和 50.0.0.1，因此，当 40.0.0.0/8 这条链路不可用时，R1 和 R2 之间的邻居关系不会受到影响，相应的 BGP 路由也不会受到影响。

总之，使用 Loopback 接口建立 BGP 邻居关系与使用物理接口来建立邻居关系相比较，前者具有更好的稳定性，且能够减少设备资源的开销。

3. 配置 EBGP 邻居

前面的实验使用物理接口来创建 R1 和 R2 的 BGP 邻居关系，配置相对简单，并且能够实现邻居关系的冗余。但是，这样的配置会产生两个 TCP 会话和两个 BGP 邻居关系，当有路由需要彼此通告时，会通过这两个邻居关系分别进行通告，因而比较消耗设备资源，并且链路的不稳定也会导致 BGP 邻居关系的不稳定。

接下来，我们将在 R2 和 R3 上使用 Loopback 0 接口来建立 EBGP 邻居关系。

```
[R2]bgp 100
[R2-bgp]peer 30.0.0.1 as-number 200

[R3]bgp 200
[R3-bgp]router-id 30.0.0.1
[R3-bgp]peer 20.0.0.1 as-number 100
```

配置完成后，在 R3 上使用 display bgp peer 命令查看 BGP 邻居关系。

```
[R3]display bgp peer
 BGP local router ID : 30.0.0.1
 Local AS number : 200
 Total number of peers : 1    Peers in established state : 0
  Peer        V    AS   MsgRcvd   MsgSent   OutQ   Up/Down     State    PrefRcv
  20.0.0.1    4    100    0         0        0     00:02:13    Idle       0
```

结果显示，R2 和 R3 的邻居状态一直停留在 Idle 状态，说明邻居关系未能正常建立。

BGP 邻居关系建立的前提条件是要能够建立起 TCP 会话，而目前 R2 和 R3 上都不存在去往对方 Loopback 0 接口的路由，因此无法建立 TCP 会话。为了解决这个问题，可以在 R2 和 R3 上配置到达对方 Loopback 0 接口的静态路由。

```
[R2]ip route-static 30.0.0.0 255.0.0.0 60.0.0.2
[R2]ip route-static 30.0.0.0 255.0.0.0 70.0.0.2

[R3]ip route-static 20.0.0.0 255.0.0.0 60.0.0.1
[R3]ip route-static 20.0.0.0 255.0.0.0 70.0.0.1
```

计算机网络实践教程

配置完成后，在 R3 上再次查看 BGP 邻居关系。

```
[R3]display bgp peer
 BGP local router ID : 30.0.0.1
 Local AS number : 200
 Total number of peers : 1    Peers in established state : 0
   Peer       V    AS    MsgRcvd  MsgSent   OutQ  Up/Down    State    PrefRcv
   20.0.0.1   4    100   0        0         0     00:12:23   Active   0
```

结果显示，R2 和 R3 之间的邻居关系一直停留在 Active 状态，说明邻居关系还是未能正常建立起来的，其原因前面已经有解释。

```
[R2]bgp 100
[R2-bgp]peer 30.0.0.1 connect-interface LoopBack 0

[R3]bgp 200
[R3-bgp]peer 20.0.0.1 connect-interface LoopBack 0
```

配置完成后，在 R3 上使用 display bgp peer 命令查看 BGP 邻居关系。

```
[R3]display bgp peer
 BGP local router ID : 30.0.0.1
 Local AS number : 200
 Total number of peers : 1    Peers in established state : 0
   Peer       V    AS    MsgRcvd  MsgSent   OutQ  Up/Down    State    PrefRcv
   20.0.0.1   4    100   0        0         0     00:00:12   Idle     0
```

结果显示，现在 R3 的邻居状态是 Idle，说明 R2 和 R3 之间的 EBGP 邻居关系仍然未能正常建立。

在默认情况下，EBGP 邻居之间在发送 BGP 报文时，TTL 值为 1，所以 EBGP 默认要求邻居之间必须物理直连。但是，当 R2 和 R3 使用 Loopback 0 接口建立邻居关系时，由于使用的不是物理直连的接口，所以 TTL 值会被多减一次，成为 0，最终 BGP 报文会被丢弃，从而导致邻居关系无法建立。为解决这一问题，可以修改 EBGP 邻居发送 BGP 报文的 TTL 值，使报文的 TTL 值大于 1。

```
[R2]bgp 100
[R2-bgp]peer 30.0.0.1 ebgp-max-hop 2

[R3]bgp 200
```

```
[R3-bgp]peer 20.0.0.1 ebgp-max-hop 2
```

配置完成后,在 R3 上查看邻居关系。

```
[R3]display bgp peer
 BGP local router ID : 30.0.0.1
 Local AS number : 200
 Total number of peers : 1    Peers in established state : 0
 Peer        V  AS   MsgRcvd  MsgSent  OutQ  Up/Down   State        PrefRcv
 20.0.0.1    4  100  4        4        0     00:02:04  Established  0
```

结果显示,R2 和 R3 之间已经建立起了 EBGP 邻居关系。

需要说明的是,在实际场景中,通常使用 Loopback 接口来建立 IBGP 邻居关系,使用物理接口建立 EBGP 邻居关系。

第 6 章 VLAN 路由配置

6.1 利用单臂路由器实现 VLAN 间路由

一、原理概述

以太网中,通常会使用 VLAN 技术隔离二层广播域来减少广播的影响,并增强网络的安全性和可管理性。其缺点是同时也严格地隔离了不同 VLAN 之间的任何二层流量,使分属于不同 VLAN 的用户不能直接互相通信。在现实中,经常会出现某些用户需要跨越 VLAN 实现通信的情况,单臂路由技术就是解决 VLAN 间通信的一种方法。

单臂路由的原理是通过一台路由器,使 VLAN 间互通数据通过路由器进行三层转发。如果在路由器上为每个 VLAN 分配一个单独的路由器物理接口,随着 VLAN 数量的增加,需要更多的接口,而路由器能提供的接口数量比较有限,所以在路由器的一个物理接口上配置子接口(即逻辑接口)的方式来实现以一当多的功能,将是一种非常好的方式。路由器同一物理接口的不同子接口作为不同 VLAN 的默认网关,当不同 VLAN 间的用户主机需要通信时,只需将数据包发送给网关,网关处理后再发送至目的主机所在 VLAN,从而实现 VLAN 间通信。由于从拓扑结构图上看,在交换机与路由器之间,数据仅通过一条物理链路传输,故被形象地称为"单臂路由"。

二、实验目的

1. 理解单臂路由的应用场景
2. 掌握路由器子接口的配置方法
3. 掌握子接口封装 VLAN 的配置方法
4. 理解单臂路由的工作原理

三、实验内容

本实验模拟公司网络场景。路由器 R1 是公司的出口网关，员工 PC 通过接入层交换机（如 SW2 和 SW3）接入公司网络，接入层交换机又通过汇聚交换机 SW1 与路由器 R1 相连。公司内部网络通过划分不同的 VLAN 隔离了不同部门之间的二层通信，保证各部门间的信息安全，但是由于业务需要，行政部、市场部和人事部之间需要能实现跨 VLAN 通信，网络管理员决定借助路由器的三层功能，通过配置单臂路由来实现。

四、实验拓扑

单臂路由器实现 VLAN 间路由配置拓扑如图 6.1 所示。

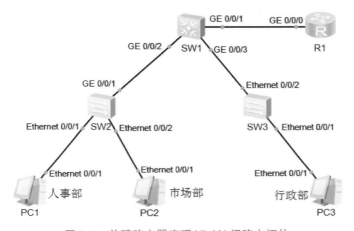

图 6.1　单臂路由器实现 VLAN 间路由拓扑

五、实验地址分配

本实验地址分配如表 6.1 所示。

表 6.1　实验 IP 地址分配表

设备	接口	IP 地址	子网掩码	默认网关	VLAN ID
R1（AR2220）	GE 0/0/0	192.168.1.1	255.255.255.0	N/A	vlan 10
	GE 0/0/0	192.168.2.1	255.255.255.0	N/A	vlan 20
	GE 0/0/0	192.168.3.1	255.255.255.0	N/A	vlan 30
PC1	Ethernet 0/0/1	192.168.1.2	255.255.255.0	192.168.1.1	vlan 10
PC2	Ethernet 0/0/1	192.168.2.2	255.255.255.0	192.168.2.1	vlan 20
PC3	Ethernet 0/0/1	192.168.3.2	255.255.255.0	192.168.3.1	vlan 30

计算机网络实践教程

六、实验步骤

1. PC 的基本配置

PC1 的基本配置如图 6.2 所示。

图 6.2　PC1 的基本配置

PC2 的基本配置如图 6.3 所示。

图 6.3　PC2 的基本配置

PC3 的基本配置如图 6.4 所示。

图 6.4　PC3 的基本配置

在 PC1 上使用 ping 命令测试属于不同 VLAN 的 PC2 或 PC3 之间的连通性。

```
PC>ping 192.168.2.2 -c 1
ping 192.168.2.2: 32 data bytes, Press Ctrl_C to break
From 192.168.1.2: Destination host unreachable
--- 192.168.1.1 ping statistics ---
 1 packet(s) transmitted
```

```
0 packet(s) received
100.00% packet loss
```

结果显示，PC1 与 PC2 之间无法正常通信，同样测试 PC1 与 PC3 之间无法正常通信。

主机发送数据包前，将会查看数据包中的目的 IP 地址，如果目的 IP 地址和本机 IP 地址在同一个网段上，主机会通过 MAC 地址直接转发到目的主机；如果目的 IP 地址和本机 IP 地址不在同一个网段上，主机只能通过 MAC 地址转发到网关。

2. 交换机的基本配置

1) SW2 的基本配置

在 SW2 上创建 vlan 10 和 vlan 20，把连接 PC1 的 Ethernet 0/0/1 和连接 PC2 的 Ethernet 0/0/2 接口配置为 Access 类型接口，并分别划分到 vlan 10 和 vlan 20 中。

```
<Huawei>system-view
[Huawei]sysname SW2
[SW2]vlan 10
[SW2-vlan10]description HR
[SW2-vlan10]vlan 20
[SW2-vlan20]description MK
[SW2-vlan20]interface Ethernet 0/0/1
[SW2-Ethernet0/0/1]port link-type access
[SW2-Ethernet0/0/1]port default vlan 10
[SW2-Ethernet0/0/1]interface Ethernet 0/0/2
[SW2-Ethernet0/0/2]port link-type access
[SW2-Ethernet0/0/2]port default vlan 20
```

2) SW3 的基本配置

在 SW3 上创建 vlan 30，把连接 PC3 的 Ethernet 0/0/1 接口配置为 Access 类型，并划分到 vlan 30。

```
<Huawei>system-view
[Huawei]sysname SW3
[SW3]vlan 30
[SW3-vlan30]description MG
[SW3-vlan30]interface Ethernet 0/0/1
[SW3-Ethernet0/0/1]port link-type access
[SW3-Ethernet0/0/1]port default vlan 30
```

交换机之间或交换机和路由器之间相连的接口需要传递多个 VLAN 信息，需要配置成 Trunk 接口。

将 SW2 和 SW3 的 GE 0/0/1 接口配置成 Trunk 类型接口，并允许所有 VLAN 通过。

```
[SW2]interface GigabitEthernet 0/0/1
[SW2-GigabitEthernet0/0/1]port link-type trunk
[SW2-GigabitEthernet0/0/1]port trunk allow-pass vlan all

[SW3]interface GigabitEthernet 0/0/1
[SW3-GigabitEthernet0/0/1]port link-type trunk
[SW3-GigabitEthernet0/0/1]port trunk allow-pass vlan all
```

3) SW1 的基本配置

在 SW1 上创建 vlan 10、vlan 20 和 vlan 30，并配置交换机和路由器相连的接口为 Trunk，允许所有 VLAN 通过。

```
<Huawei>system-view
[Huawei]sysname SW1
[SW1]vlan batch 10 20 30
[SW1]interface GigabitEthernet 0/0/2
[SW1-GigabitEthernet0/0/2]port link-type trunk
[SW1-GigabitEthernet0/0/2]port trunk allow-pass vlan all
[SW1-GigabitEthernet0/0/2]interface GigabitEthernet 0/0/3
[SW1-GigabitEthernet0/0/3]port link-type trunk
[SW1-GigabitEthernet0/0/3]port trunk allow-pass vlan all
[SW1-GigabitEthernet0/0/3]interface GigabitEthernet 0/0/1
[SW1-GigabitEthernet0/0/1]port link-type trunk
[SW1-GigabitEthernet0/0/1]port trunk allow-pass vlan all
```

3. 配置路由器子接口和 IP 地址

由于路由器 R1 只有一个实际的物理接口与交换机 SW1 相连，可以在路由器上配置不同的逻辑子接口来作为不同 VLAN 的网关，从而达到节省路由器接口的目的。

在 R1 上创建子接口 GE 0/0/0.1，配置 IP 地址 192.168.1.1/24，作为人事部网关地址；创建子接口 GE 0/0/0.2，配置 IP 地址 192.168.2.1/24，作为市场部网关地址；创建子接口 GE 0/0/0.3，配置 IP 地址 192.168.3.1/24，作为行政部网关地址。

```
<Huawei>system-view
[Huawei]sysname R1
[R1]interface GigabitEthernet 0/0/0.1
[R1-GigabitEthernet0/0/0.1]ip address 192.168.1.1 255.255.255.0
[R1-GigabitEthernet0/0/0.1]interface GigabitEthernet 0/0/0.2
[R1-GigabitEthernet0/0/0.2]ip address 192.168.2.1 255.255.255.0
[R1-GigabitEthernet0/0/0.2]interface GigabitEthernet 0/0/0.3
[R1-GigabitEthernet0/0/0.3]ip address 192.168.3.1 255.255.255.0
```

第 6 章　VLAN 路由配置

在 PC1 上使用 ping 命令再次测试与 PC2 和 PC3 间的连通性。

```
PC>ping 192.168.2.2 -c 1
ping 192.168.2.2: 32 data bytes, Press Ctrl_C to break
From 192.168.1.2: Destination host unreachable
--- 192.168.1.1 ping statistics ---
 1 packet(s) transmitted
 0 packet(s) received
 100.00% packet loss

PC>ping 192.168.3.2 -c 1
ping 192.168.3.2: 32 data bytes, Press Ctrl_C to break
From 192.168.1.2: Destination host unreachable
--- 192.168.1.1 ping statistics ---
 1 packet(s) transmitted
 0 packet(s) received
 100.00% packet loss
```

结果显示，通信仍然无法建立。

4. 配置路由器子接口封装 VLAN

虽然目前已经创建了不同的子接口，并配置了相关 IP 地址，但是仍然无法通信。这是由于处于不同 VLAN 下，不同网段的 PC 间要实现互相通信，数据包必须通过路由器进行中转。由 SW1 发送到 R1 的数据都加上了 VLAN 标签，而路由器作为三层设备，默认无法处理带了 VLAN 标签的数据包。因此，需要在路由器上的子接口下配置对应 VLAN 的封装，使路由器能够识别和处理 VLAN 标签，包括剥离和封装 VLAN 标签。

在 R1 的子接口 GE 0/0/0.1 上封装 vlan 10，在 R1 的子接口 GE 0/0/0.2 上封装 vlan 20，在 R1 的子接口 GE 0/0/0.3 上封装 vlan 30，并开启子接口的 ARP 广播功能。

使用 dot1q termination vid 命令配置子接口对一层 tag 报文的终结功能。即配置该命令后，路由器子接口在接收带有 VLAN tag 的报文时，将剥掉 tag 进行三层转发，在发送报文时，会将与该子接口对应 VLAN 的 VLAN tag 添加到报文中。使用 arp broadcast enable 命令开启子接口的 ARP 广播功能。如果不配置该命令，则会导致该子接口无法主动发送 ARP 广播报文，以及向外转发 IP 报文。

```
[R1]interface GigabitEthernet 0/0/0.1
[R1-GigabitEthernet0/0/0.1]dot1q termination vid 10
[R1-GigabitEthernet0/0/0.1]arp broadcast enable
[R1-GigabitEthernet0/0/0.1]interface GigabitEthernet 0/0/0.2
[R1-GigabitEthernet0/0/0.2]dot1q termination vid 20
[R1-GigabitEthernet0/0/0.2]arp broadcast enable
[R1-GigabitEthernet0/0/0.2]interface GigabitEthernet 0/0/0.3
```

```
[R1-GigabitEthernet0/0/0.3]dot1q termination vid 30
[R1-GigabitEthernet0/0/0.3]arp broadcast enable
```

配置完成后，在 R1 上使用 display ip interface brief 命令查看接口状态。

```
[R1]display ip interface brief
*down: administratively down
^down: standby
......
Interface                 IP Address/Mask      Physical   Protocol
GigabitEthernet0/0/0      unassigned           up         down
GigabitEthernet0/0/0.1    192.168.1.1/24       up         up
GigabitEthernet0/0/0.2    192.168.2.1/24       up         up
GigabitEthernet0/0/0.3    192.168.3.1/24       up         up
......
```

结果显示，3 个子接口的物理状态和协议状态都正常。

在路由器 R1 上使用 display ip routing-table 命令查看 R1 的路由表。

```
[R1]display ip routing-table
Route Flags: R - relay, D - download to fib
------------------------------------------------------------------------
Routing Tables: Public
        Destinations : 13      Routes : 13
Destination/Mask   Proto   Pre  Cost  Flags  NextHop      Interface
   ......
192.168.1.0/24    Direct   0    0     D      192.168.1.1  GigabitEthernet0/0/0.1
192.168.1.1/32    Direct   0    0     D      127.0.0.1    GigabitEthernet0/0/0.1
192.168.2.0/24    Direct   0    0     D      192.168.2.1  GigabitEthernet0/0/0.2
192.168.2.1/32    Direct   0    0     D      127.0.0.1    GigabitEthernet0/0/0.2
192.168.3.0/24    Direct   0    0     D      192.168.3.1  GigabitEthernet0/0/0.3
   ......
```

结果显示，路由表中已经有了 192.168.1.0/24、192.168.2.0/24 和 192.168.3.0/24 的路由条目，并且都是路由器 R1 的直连路由，类似于路由器上的直连物理接口。

在 PC1 上使用 ping 命令分别测试与网关和 PC2 之间的连通性。

```
PC>ping 192.168.1.1 -c 1
ping 192.168.1.1: 32 data bytes, Press Ctrl_C to break
From 192.168.1.1: bytes=32 seq=1 ttl=255 time=63 ms
......
PC>ping 192.168.2.2 -c 1
```

第 6 章　VLAN 路由配置

```
ping 192.168.2.2: 32 data bytes, Press Ctrl_C to break
From 192.168.2.2: bytes=32 seq=1 ttl=127 time=109 ms
……
```

结果显示，通信正常。在 PC1 上使用 **tracert** 命令查看到达 PC2 的路径。

```
PC>tracert 192.168.2.2
traceroute to 192.168.2.2, 8 hops max
(ICMP), press Ctrl+C to stop
 1 192.168.1.1      62 ms    62 ms    63 ms
 2 *192.168.2.2    109 ms   125 ms
```

结果显示 PC1 先把 ICMP 包发送给自己的网关 192.168.1.1，然后再由网关发送到 PC2。现以 PC1 ping PC2 为例，分析单臂路由的整个运作过程。

两台 PC 由于处于不同的网络中，这时 PC1 会将数据包发往自己的网关，即路由器 R1 的子接口 GE 0/0/0.1 的地址 192.168.1.1。

数据包到达路由器 R1 后，由于路由器的子接口 GE 0/0/0.1 已经配置了 VLAN 封装，当接收到 PC1 发送的 vlan 10 的数据帧时，发现数据帧的 VLAN ID 与自身 GE 0/0/0.1 接口配置的 VLAN ID 一样，便会剥离掉数据帧的 VLAN 标签后通过三层路由转发。

通过查找路由表后，发现数据包中的目的地址 192.168.2.2 所属的 192.168.2.0/24 网段的路由条目，已经是 R1 上的直连路由，且出接口为 GE 0/0/0.2，便将该数据包发送至 GE 0/0/0.2 接口。当 GE 0/0/0.2 接口接收到一个没有带 VLAN 标签的数据帧时，便会加上自身接口所配置的 vlan 20 后再进行转发，然后通过交换机将数据帧顺利转发给 PC2。

6.2　利用三层交换机实现 VLAN 间路由

一、原理概述

三层交换机在二层交换机的基础上增加了三层的路由功能，很好地解决了单臂路由器实现 VLAN 间路由带来的不便，为网络设计提供了一个高效的解决方案。

vlanif 接口是基于网络层的接口，可以配置 IP 地址。借助 vlanif 接口，三层交换机就能实现路由转发功能。

二、实验目的

1. 掌握配置 vlanif 接口的方法

2. 理解数据包跨 VLAN 路由的原理
3. 掌握测试多层交换网络连通性的方法

三、实验内容

本实验模拟企业网络场景。公司有行政部、人事部和市场部三个部门，分别规划 vlan 10、vlan 20 和 vlan 30。其中行政部有两台主机 PC1 和 PC2，人事部有一台主机 PC3，市场部有一台主机 PC4。所有主机通过三层交换机 SW1 相连。网络管理员通过三层交换机实现 3 个 VLAN 之间互相访问。

四、实验拓扑

三层交换机实现 VLAN 间路由拓扑如图 6.5 所示。

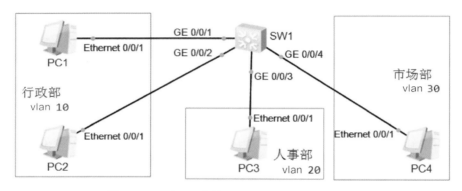

图 6.5　利用三层交换机实现 VLAN 间路由拓扑

五、实验地址分配

本实验地址分配如表 6.2 所示。

表 6.2　实验 IP 地址分配表

设备	接口	IP 地址	子网掩码	默认网关	VLAN ID
PC1	Ethernet 0/0/1	192.168.1.2	255.255.255.0	192.168.1.1	vlan 10
PC2	Ethernet 0/0/1	192.168.1.3	255.255.255.0	192.168.1.1	vlan 10
PC3	Ethernet 0/0/1	192.168.2.2	255.255.255.0	192.168.2.1	vlan 20
PC4	Ethernet 0/0/1	192.168.3.2	255.255.255.0	192.168.3.1	vlan 30

续表

设备	接口	IP 地址	子网掩码	默认网关	VLAN ID
SW1 （S5700）	vlanif 10	192.168.1.1	255.255.255.0		
	vlanif 20	192.168.2.1	255.255.255.0		
	vlanif 30	192.168.3.1	255.255.255.0		

六、实验步骤

1. PC 的基本配置

PC1 的基本配置如图 6.6 所示。

图 6.6 PC1 的基本配置

PC2 的基本配置如图 6.7 所示。

图 6.7 PC2 的基本配置

PC3 的基本配置如图 6.8 所示。

图 6.8 PC3 的基本配置

PC4 的基本配置如图 6.9 所示。

图 6.9　PC4 的基本配置

在 PC1 上使用 ping 命令测试属同一个 VLAN 的 PC2 之间的连通性。

```
PC>ping 192.168.1.3 -c 1
ping 192.168.1.3: 32 data bytes, Press Ctrl_C to break
From 192.168.1.3: bytes=32 seq=1 ttl=128 time=31 ms
--- 192.168.1.3 ping statistics ---
……
  0.00% packet loss
```

结果显示，通信正常。

在 PC1 上使用 ping 命令测试属于不同 VLAN 的 PC3 或 PC4 之间的连通性。

```
PC>ping 192.168.2.2 -c 1
ping 192.168.2.2: 32 data bytes, Press Ctrl_C to break
From 192.168.1.2: Destination host unreachable
--- 192.168.1.1 ping statistics ---
……
  100.00% packet loss
```

PC1 与 PC3 之间无法正常通信，同样测试 PC1 与 PC4 之间无法正常通信。

主机发送数据包前，将会查看数据包中的目的 IP 地址，如果目的 IP 地址和本机 IP 地址在同一个网段上，主机会通过 MAC 地址直接转发到目的主机；如果目的 IP 地址和本机 IP 地址不在同一个网段上，主机只能通过 MAC 地址转发到网关。

2. 交换机的基本配置

在三层交换机 SW1 上创建 vlan 10、vlan 20 和 vlan 30，把行政部的 PC1 和 PC2 划入 vlan 10 中，把人事部的 PC3 划入 vlan 20，把市场部的 PC4 划入 vlan 30。

```
<Huawei>system-view
[Huawei]sysname SW1
[SW1]vlan batch 10 20 30
[SW1-vlan30]interface GigabitEthernet 0/0/1
[SW1-GigabitEthernet0/0/1]port link-type access
```

```
[SW1-GigabitEthernet0/0/1]port default vlan 10
[SW1-GigabitEthernet0/0/1]interface GigabitEthernet 0/0/2
[SW1-GigabitEthernet0/0/2]port link-type access
[SW1-GigabitEthernet0/0/2]port default vlan 10
[SW1-GigabitEthernet0/0/2]interface GigabitEthernet 0/0/3
[SW1-GigabitEthernet0/0/3]port link-type access
[SW1-GigabitEthernet0/0/3]port default vlan 20
[SW1-GigabitEthernet0/0/3]interface GigabitEthernet 0/0/4
[SW1-GigabitEthernet0/0/4]port link-type access
[SW1-GigabitEthernet0/0/4]port default vlan 30
```

3. 配置三层交换机实现 VLAN 间通信

现在需要通过 VLAN 间路由来实现通信,在三层交换机上配置 vlanif 接口。

在 SW1 上使用 interface vlanif 命令创建 vlanif 接口,指定 vlanif 接口所对应的 vlan ID 为 10,并进入 vlanif 接口视图,在接口视图下配置 IP 地址 192.168.1.1/24;创建对应 vlan 20 的 vlanif 接口,地址配置为 192.168.2.1/24;创建对应 vlan 30 的 vlanif 接口,地址配置为 192.168.3.1/24。

```
[SW1]interface vlanif 10
[SW1-vlanif10]ip address 192.168.1.1 24
[SW1-vlanif10]interface vlanif 20
[SW1-vlanif20]ip address 192.168.2.1 24
[SW1-vlanif20]interface vlanif 30
[SW1-vlanif30]ip address 192.168.3.1 24
```

配置完成后,查看接口状态。

```
[SW1]display ip interface brief
*down: administratively down
……
Interface        IP Address/Mask      Physical    Protocol
MEth0/0/1        unassigned           down        down
NULL0            unassigned           up          up(s)
vlanif1          unassigned           down        down
vlanif10         192.168.1.1/24       up          up
vlanif20         192.168.2.1/24       up          up
vlanif30         192.168.3.1/24       up          up
```

结果显示,3 个 vlanif 接口已经生效。再次使用 ping 命令测试 PC1 与 PC3 之间的连通性。

```
PC>ping 192.168.2.2 -c 1
ping 192.168.2.2: 32 data bytes, Press Ctrl_C to break
From 192.168.2.2: bytes=32 seq=1 ttl=127 time=62 ms
```

```
--- 192.168.2.2 ping statistics ---
 1 packet(s) transmitted
 1 packet(s) received
 0.00% packet loss
```

通信正常,实现了行政部与人事部主机间的通信。同样测试其他主机之间的连通性。

在 PC1 上查看 ARP 信息。

```
PC>arp -a
Internet Address     Physical Address      Type
192.168.1.1          4C-1F-CC-F0-48-E9     dynamic
```

结果显示,目前 PC 上 ARP 解析的地址只有交换机 vlanif 10 的地址,而没有对端的地址,PC1 先将数据包发送至网关,即对应的 vlanif 10 接口,再由网关转发到对端。

第 7 章 广域网的配置

7.1 广域网接入配置

一、原理概述

如果要将本地网络与远程网络连接起来,有时要用到广域网接入服务。目前广域网技术包括高级数据链路控制 HDLC(high-level data link control)、点到点协议 PPP(point to point protocol)等协议。

HDLC 是由 ISO 制定的,是通信领域曾广泛应用的一个数据链路层协议。但是随着技术进步,目前通信信道的可靠性得到保障,HDLC 在公网的应用逐渐消失。

PPP 是一种数据链路层协议,主要用来进行点到点之间的数据传输。PPP 设计初衷是为两个对等节点之间的 IP 流量提供一种封装协议,它是在串行线 IP 协议 SLIP(serial line IP)的基础上发展而来的。由于 SLIP 只支持异步传输方式、无协商过程、只能承载 IP 一种网络报文等问题,在发展过程中,逐步被 PPP 协议所替代。PPP 与 HDLC 的主要区别是:HDLC 是面向位的,而 PPP 是面向字节的。PPP 是一种多协议成帧机制,适用于调制解调器。

二、实验目的

1. 掌握 PPP 的基本配置
2. 掌握 HDLC 的基本配置
3. 理解 PPP 与 HDLC 的异同

三、实验内容

本实验模拟某公司内联网的场景。总公司、分公司 A 和分公司 B 通过路由器 R1、R2 和 R3 接入某 ISP 的广域网。R2 与 R1 之间链路为串行链路,封装 PPP 协议;R3 与 R1 之间链路为串行链路，封装 HDLC 协议。配置路由表，使各分公司与总公司之间能互相访问。

四、实验拓扑

广域网接入配置拓扑如图 7.1 所示。

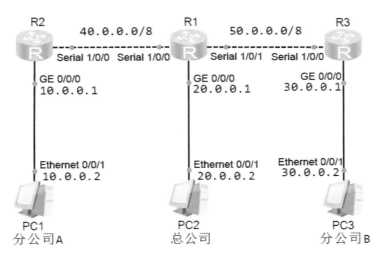

图 7.1 广域网接入配置拓扑

五、实验地址分配

本实验地址分配如表 7.1 所示。

表 7.1 实验 IP 地址分配表

设备	接口	IP 地址	子网掩码	默认网关
PC1	Ethernet 0/0/1	10.0.0.2	255.0.0.0	10.0.0.1
PC2	Ethernet 0/0/1	20.0.0.2	255.0.0.0	20.0.0.1
PC3	Ethernet 0/0/1	30.0.0.2	255.0.0.0	30.0.0.1
R1（AR2220）	GE 0/0/0	20.0.0.1	255.0.0.0	N/A
R1（AR2220）	Serial 1/0/0	40.0.0.2	255.0.0.0	N/A
R1（AR2220）	Serial 1/0/1	50.0.0.1	255.0.0.0	N/A
R2（AR2220）	GE 0/0/0	10.0.0.1	255.0.0.0	N/A
R2（AR2220）	Serial 1/0/0	40.0.0.1	255.0.0.0	N/A
R3（AR2220）	GE 0/0/0	30.0.0.1	255.0.0.0	N/A
R3（AR2220）	Serial 1/0/0	50.0.0.2	255.0.0.0	N/A

六、实验步骤

1. PC 的基本配置

PC1 的基本配置如图 7.2 所示。

图 7.2　PC1 的基本配置

PC2 的基本配置如图 7.3 所示。

图 7.3　PC2 的基本配置

PC3 的基本配置如图 7.4 所示。

图 7.4　PC3 的基本配置

2. 路由器的基本配置

根据实验地址分配表完成 R1、R2 和 R3 的基本配置。

1) R1 的配置

```
<Huawei>system-view
[Huawei]sysname R1
[R1]interface GigabitEthernet 0/0/0
[R1-GigabitEthernet0/0/0]ip address 20.0.0.1 255.0.0.0
```

```
[R1-GigabitEthernet0/0/0]interface serial 1/0/0
[R1-Serial1/0/0]ip address 40.0.0.2 255.0.0.0
[R1-Serial1/0/0]interface serial 1/0/1
[R1-Serial1/0/1]ip address 50.0.0.1 255.0.0.0
```

2) R2 的配置

```
<Huawei>system-view
[Huawei]sysname R2
[R2]interface GigabitEthernet 0/0/0
[R2-GigabitEthernet0/0/0]ip address 10.0.0.1 255.0.0.0
[R2-GigabitEthernet0/0/0]interface serial 1/0/0
[R2-Serial1/0/0]ip address 40.0.0.1 255.0.0.0
```

3) R3 的配置

```
<Huawei>system-view
[Huawei]sysname R3
[R3]interface GigabitEthernet 0/0/0
[R3-GigabitEthernet0/0/0]ip address 30.0.0.1 255.0.0.0
[R3-GigabitEthernet0/0/0]interface serial 1/0/0
[R3-Serial1/0/0]ip address 50.0.0.2 255.0.0.0
```

在 R1 上使用 ping 命令检测与 R2 之间到连通性。

```
[R1]ping -c 1 40.0.0.1
 ping 40.0.0.1: 56  data bytes, press CTRL_C to break
   Reply from 40.0.0.1: bytes=56 Sequence=1 ttl=255 time=60 ms
 --- 40.0.0.1 ping statistics ---
   1 packet(s) transmitted
   1 packet(s) received
   0.00% packet loss
```

结果显示，R1 与直连的 R2 之间正常通信，同样方法测试其余直连网段的连通性。

3. 配置 PPP

默认情况下，串行接口封装的链路层协议即为 PPP，所以一般不需要修改链路层协议。可以直接在 R1 上使用 display interface serial 1/0/0 命令进行查看。

```
[R1]display interface serial 1/0/0
Serial1/0/0 current state : UP
```

```
……
Internet Address is 40.0.0.2/8
Link layer protocol is PPP
LCP opened, IPCP opened
……
```

也可以通过 link-protocol ppp 命令设置路由器接口链路层协议。

```
[R1-Serial1/0/0]link-protocol ppp

[R2-Serial1/0/0]link-protocol ppp
```

PPP 协议在建立连接时可以选择认证方式，R1 作为认证方，用户信息保存在本地，要求 R2 对其进行 PAP/CHAP 认证。在路由器 R1 上创建本地用户及域并配置端口 PPP 认证方式为 PAP/CHAP，认证域为 test。

```
[R1]aaa
[R1-aaa]local-user hbeutc@hbeu password cipher xxgcx
[R1-aaa]local-user hbeutc@hbeu service-type ppp
[R1-aaa]authentication-scheme system-a
[R1-aaa-authen-system-a]quit
[R1-aaa]domain test
[R1-aaa-domain-test]authentication-scheme system-a
[R1]interface serial 1/0/0
[R1-Serial1/0/0]ppp authentication-mode pap domain test
```

如果使用 CHAP 方式认证，以上接口配置为：

```
[R1]interface serial 1/0/0
[R1-Serial1/0/0]ppp authentication-mode pap domain test
```

在路由器 R2 上配置本地被 R1 要求验证时需要发送的用户名和密码。

```
[R2]interface serial 1/0/0
[R2-Serial1/0/0]ppp pap local-user hbeutc@hbeu password simple xxgcx
```

或者：

```
[R2]interface serial 1/0/0
[R2-Serial1/0/0]ppp chap local-user hbeutc@hbeu password simple xxgcx
```

在路由器 R1 上通过命令 display interface 命令查看接口的配置信息，端口的物理层和数据链路层的状态都是 UP 状态，并且 PPP 的 LCP 和 IPCP 都是 opened 状态，说明链路的 PPP 协商已经成功。

```
[R1]display interface serial1/0/0
Serial1/0/0 current state : UP
……
Internet Address is 40.0.0.2/8
Link layer protocol is PPP
LCP opened, IPCP opened
……
```

4. 配置 HDLC

在 R1 的 Serial 1/0/1 和 R3 的 Serial 1/0/0 接口上分别使用 link-protocol 命令配置链路层协议为 HDLC。

```
[R1]interface serial 1/0/1
[R1-Serial1/0/1]link-protocol hdlc
 Warning:The encapsulation protocol of the link will be
changed.Continue?[Y/N]:y
[R3]interface serial 1/0/0
[R3-Serial1/0/0]link-protocol hdlc
 Warning:The encapsulation protocol of the link will be
changed.Continue?[Y/N]:y
```

5. 配置路由表

在 R1 上配置静态路由，R2 和 R3 上配置默认路由。

```
[R1]ip route-static 10.0.0.0 255.0.0.0 40.0.0.1
[R1]ip route-static 30.0.0.0 255.0.0.0 50.0.0.2

[R2]ip route-static 0.0.0.0 0.0.0.0 40.0.0.2

[R3]ip route-static 0.0.0.0 0.0.0.0 50.0.0.1
```

6. 测试

在 PC1 上使用 ping 命令测试与 PC3 间的连通性。结果显示，可以正常通信。

```
PC>ping 30.0.0.2 -c 1
ping 30.0.0.2: 32 data bytes, Press Ctrl_C to break
From 30.0.0.2: bytes=32 seq=1 ttl=125 time=31 ms
--- 30.0.0.2 ping statistics ---
 1 packet(s) transmitted
 1 packet(s) received
 0.00% packet loss
```

7.2 帧中继配置

一、原理概述

帧中继是一种有效的数据传输技术，它可以在一对一或者一对多的应用中快速而低廉地传输数字信息。它可以应用于语音、数据通信，既可用于广域网也可用于局域网的通信。每个帧中继用户将得到一个接到帧中继节点的专线。帧中继网络对于端用户来说，它通过一条经常改变且对用户不可见的信道来处理和其他用户间的数据传输。

帧中继网络的端设备用虚电路来连接。每条虚电路用数据链路连接标识符（data link connection identifier，DLCI）定义的一条帧中继连接通道，提供了用户设备（如路由器和主机）之间进行数据通信的能力。

DLCI 用来标识各端点的一个具有局部意义的数值。多个永久虚电路（permanent virtual circuit，PVC）可以连接到同一个物理终端。

逆向地址解析协议（inverse ARP，IARP）用于帧中继网络中自动建立路由器 IP 地址与帧中继 DLCI 的映射关系。IARP 是一种在网络中建立动态路由的方法，让接入路由器知道与虚电路相关联的对端设备的网络地址。IARP 用于在帧中继网络中自动建立路由器 IP 地址与帧中继 DLCI 的映射关系。

二、实验目的

1. 掌握帧中继的基本配置
2. 掌握静态和动态映射的配置

三、实验内容

本实验模拟企业网络场景。公司总部和分公司分别设在不同地方，总部路由器 R1 和分公司路由器 R2 通过帧中继网络相连，通过帧中继协议方式互连 R1 和 R2，实现 IP 层互通。

四、实验拓扑

帧中继配置拓扑如图 7.5 所示。

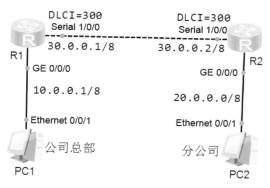

图 7.5 帧中继配置拓扑

五、实验地址分配

本实验 IP 地址分配如表 7.2 所示。

表 7.2 IP 地址分配表

设备	接口	IP 地址	子网掩码	默认网关	DLCI
PC1	Ethernet 0/0/1	10.0.0.2	255.0.0.0	10.0.0.1	N/A
PC2	Ethernet 0/0/1	20.0.0.2	255.0.0.0	20.0.0.1	N/A
R1（AR2220）	GE 0/0/0	10.0.0.1	255.0.0.0	N/A	N/A
	Serial 1/0/0	30.0.0.1	255.0.0.0	N/A	400
R2（AR2220）	GE 0/0/0	20.0.0.1	255.0.0.0	N/A	N/A
	Serial 1/0/0	30.0.0.2	255.0.0.0	N/A	400

六、实验步骤

1. PC 的基本配置

PC1 的基本配置如图 7.6 所示。

第 7 章 广域网的配置

图 7.6 PC1 的基本配置

PC2 的基本配置如图 7.7 所示。

图 7.7 PC2 的基本配置

2. 路由器的基本配置

1) R1 的配置

```
<Huawei>system-view
[Huawei]sysname R1
[R1]interface GigabitEthernet 0/0/0
[R1-GigabitEthernet0/0/0]ip address 10.0.0.1 255.0.0.0
[R1-GigabitEthernet0/0/0]interface serial 1/0/0
[R1-Serial1/0/0]ip add 30.0.0.1 255.0.0.0
```

2) R2 的配置

```
<Huawei>system-view
[Huawei]sysname R2
[R2]interface GigabitEthernet 0/0/0
[R2-GigabitEthernet0/0/0]ip address 20.0.0.1 255.0.0.0
[R2-GigabitEthernet0/0/0]interface serial 1/0/0
[R2-Serial1/0/0]ip address 30.0.0.2 255.0.0.0
```

3. 静态与动态映射的配置

帧中继接口在转发数据包时必须查找帧中继地址映射表来确定下一跳的 DLCI。地址映射表中存放对端 IP 地址和下一跳的 DLCI 的映射关系。只有找到相应的映射表项，才能完成二层帧中继帧头的封装，这个机制类似于以太网中 ARP 机制。该地址映射表可以手动（静态）配置，也可以使用 IARP 协议来自动（动态）配置。

公司总部使用静态映射，在 R1 的 Serial 1/0/0 接口配置链路层协议为 FR，端口类型为 DTE，并使用 **fr map ip** 命令手工配置 R2 的 IP 地址与 DLCI 的静态映射。

```
[R1]interface serial 1/0/0
[R1-Serial1/0/0]link-protocol fr
Warning:The encapsulation protocol of the link will be changed.Continue?[Y/N]:y
[R1-Serial1/0/0]fr interface-type dte
[R1-Serial1/0/0]fr dlci 300
[R1-Serial1/0/0]undo fr inarp
[R1-Serial1/0/0]fr map ip 30.0.0.2 300
```

分公司使用静态映射，在 R2 的 Serial 1/0/0 接口配置链路层协议为 FR，端口类型为 DCE，关闭逆向地址解析功能，使用 **fr map ip** 命令手工配置 R1 的 IP 地址与 DLCI 的静态映射。将 R1 的 IP 地址 30.0.0.1 与 R2 本端 DLCI 300 配置为一条静态地址映射，即 R2 通过下一跳 DLCI 300 来访问 R1。

```
[R2]interface serial 1/0/0
[R2-Serial1/0/0]link-protocol fr
Warning: The encapsulation protocol of the link will be changed. Continue?[Y/N]:y
[R2-Serial1/0/0]fr interface-type dce
[R2-Serial1/0/0]undo fr inarp
[R2-Serial1/0/0]fr dlci 300
[R2-fr-dlci-Serial1/0/0-300]quit
[R2-Serial1/0/0]fr map ip 30.0.0.1 300
```

帧中继接口的逆向地址解析功能是默认开启的，此时，fr inarp 命令可以不配置。也可以在上述静态配置基础上使用 fr inarp 命令启动动态配置，以 R2 为例。

```
[R2]interface serial 1/0/0
[R2-Serial1/0/0]link-protocol fr
Warning: The encapsulation protocol of the link will be changed. Continue?[Y/N]:y
[R2-Serial1/0/0]fr interface-type dce
[R2-Serial1/0/0]fr dlci 300
[R2-fr-dlci-Serial1/0/0-300]quit
[R2-Serial1/0/0]fr inarp
```

默认情况下，帧中继不支持广播或组播数据的转发。如果需要在帧中继上运行一些动态路由协议，如 OSPF 协议，需要在静态映射后面添加 broadcast 参数，从而使 PVC 能够正常发送来自路由协议的广播或组播数据。

```
[R2]interface serial 1/0/0
[R2-Serial1/0/0]fr map ip 30.0.0.1 300 broadcast
```

配置完成后，在 R1 和 R2 上使用 display fr map-info 命令查看 PVC 的建立情况。

```
[R1]display fr map-info
Map Statistics for interface Serial1/0/0 (DTE)
 DLCI = 300, IP 30.0.0.2, Serial1/0/0
  create time = 2022/07/28 16:05:33, status = ACTIVE
  encapsulation = ietf, vlink = 2

[R2]display fr map-info
Map Statistics for interface Serial1/0/0 (DCE)
 DLCI = 300, IP 30.0.0.1, Serial1/0/0
  create time = 2022/07/28 15:01:42, status = ACTIVE
  encapsulation = ietf, vlink = 3
```

结果显示，R1 和 R2 上分别有一条 PVC，且均为激活状态。
在 R1 上使用 ping 命令测试 R1 与 R2 之间的连通性，通信正常。

```
<R1>ping -c 1 30.0.0.2
  ping 30.0.0.2: 56  data bytes, press CTRL_C to break
    Reply from 30.0.0.2: bytes=56 Sequence=1 ttl=255 time=20 ms
  --- 30.0.0.2 ping statistics ---
    1 packet(s) transmitted
    1 packet(s) received
    0.00% packet loss
    round-trip min/avg/max = 20/20/20 ms
```

在 R1 上使用 ping 命令测试 PC1 与 PC2 之间的连通性，无法正常通信。

4．配置路由表

为了使 PC1 与 PC2 能互相通信，在 R1 和 R2 上配置静态路由。

```
[R1]ip route-static 20.0.0.0 255.0.0.0 30.0.0.2
```

```
[R1]ip route-static 10.0.0.0 255.0.0.0 30.0.0.1
```

5. 测试

在 PC1 上使用 ping 命令测试与 PC2 之间的连通性,通信正常。

```
PC>ping 20.0.0.2 -c 1
ping 20.0.0.2: 32 data bytes, Press Ctrl_C to break
From 20.0.0.2: bytes=32 seq=1 ttl=126 time=15 ms
--- 20.0.0.2 ping statistics ---
 1 packet(s) transmitted
 1 packet(s) received
 0.00% packet loss
 round-trip min/avg/max = 15/15/15 ms
```

第8章 访问控制列表 ACL 配置

8.1 基本 ACL 配置

一、原理概述

访问控制列表（access contrl list，ACL）是应用在路由器接口的指令列表（规则），这些指令列表用来告诉路由器，哪些数据包可以通过，哪些数据包需要拒绝。

ACL 由 permit 和 deny 语句组成的一系列有顺序的规则集合，这些规则根据数据包的源地址、目的地址、源端口、目的端口等信息来描述。

按照访问控制列表的用途，ACL 可以分为 4 类，即基本 ACL、高级 ACL、基于接口 ACL、基于 MAC 地址 ACL。基本 ACL 可使用报文的源 IP 地址、时间段来定义规则。

按照访问控制列表的使用用途是依靠数字的范围来指定的，范围为 1000~1999 的数字型访问控制列表是基于接口 ACL，范围为 2000~2999 的数字型 ACL 是基本 ACL，范围为 3000~3999 的数字型是高级 ACL，范围为 4000~4999 的数字型 ACL 是基于 MAC 地址 ACL。

一个 ACL 可以由多条 "deny/permit" 语句组成，每一条语句描述一条规则，每条规则有一个 Rule-ID。Rule-ID 可以由用户进行定义，也可以由系统自动根据步长生成，默认步长为 5，Rule-ID 默认按照配置先后顺序分配 0、5、10、15 等，匹配顺序按照 ACL 的 Rule-ID 的顺序，从小到大进行匹配。建议用户自定义时也按照默认步长 5 进行设置，这样做可以给增加规则留有空间。

一个访问控制列表的实现可以分为三个步骤：访问控制列表的创建、基本访问控制列表的规则制定、将具体的 ACL 应用到相应接口上。

二、实验目的

1. 理解基本访问控制列表的应用场景
2. 掌握配置基本访问控制列表的方法

三、实验内容

本实验模拟企业网络环境,路由器 Router3 为企业核心路由器,路由器 Router1 为分公司某部门的网关,路由器 Router2 为分公司通往总部网关设备。需求分析是在 Router2 过滤分公司某部门的访问。

四、实验拓扑

基本 ACL 配置拓扑如图 8.1 所示。

图 8.1 基本 ACL 配置拓扑

五、实验地址分配

本实验地址分配如表 8.1 所示。

表 8.1 实验地址分配

设备	接口	IP 地址	子网掩码	默认网关
Router1（AR2220）	GE 0/0/0	20.0.0.1	255.0.0.0	N/A
	Loopback 0	1.1.1.1	255.255.255.255	N/A
Router2（AR2220）	GE 0/0/0	20.0.0.2	255.0.0.0	N/A
	GE 0/0/1	30.0.0.1	255.0.0.0	N/A
	Loopback 0	2.2.2.2	255.255.255.255	N/A
Router3（AR2220）	GE 0/0/0	30.0.0.2	255.0.0.0	N/A
	Loopback 0	3.3.3.3	255.255.255.255	N/A

六、实验步骤

1. 基本配置

根据实验地址分配表进行相应的基本配置,并使用 ping 命令测试各直连线的连通性。

1) Router1 的基本配置

```
<Huawei>system-view
[Huawei]sysname Router1
[Router1]interface GigabitEthernet 0/0/0
```

```
[Router1-GigabitEthernet0/0/0]ip address 20.0.0.1 8
[Router1-GigabitEthernet0/0/0]interface loopback 0
[Router1-LoopBack0]ip address 1.1.1.1 32
```

2) Router2 的基本配置

```
<Huawei>system-view
[Huawei]sysname Router2
[Router2]interface GigabitEthernet 0/0/0
[Router2-GigabitEthernet0/0/0]ip address 20.0.0.2 8
[Router2-GigabitEthernet0/0/0]interface GigabitEthernet 0/0/1
[Router2-GigabitEthernet0/0/1]ip address 30.0.0.1 8
[Router2-GigabitEthernet0/0/1]interface loopback 0
[Router2-LoopBack0]ip address 2.2.2.2 32
```

3) Router3 的基本配置

```
<Huawei>system-view
[Huawei]sysname Router3
[Router3]interface GigabitEthernet 0/0/0
[Router3-GigabitEthernet0/0/0]ip address 30.0.0.2 8
[Router3-GigabitEthernet0/0/0]interface loopback 0
[Router3-LoopBack0]ip address 3.3.3.3 32
```

2. OSPF 配置

路由器基本配置完成后，使用 OSPF 配置动态路由表。

```
[Router1]ospf 1
[Router1-ospf-1]area 0
[Router1-ospf-1-area-0.0.0.0]network 20.0.0.0 0.255.255.255
[Router1-ospf-1-area-0.0.0.0]network 1.1.1.1 0.0.0.0
------------------------------------------------------------
[Router2]ospf 1
[Router2-ospf-1]area 0
[Router2-ospf-1-area-0.0.0.0]network 20.0.0.0 0.255.255.255
[Router2-ospf-1-area-0.0.0.0]network 30.0.0.0 0.255.255.255
[Router2-ospf-1-area-0.0.0.0]network 2.2.2.2 0.0.0.0
------------------------------------------------------------
[Router3]ospf 1
[Router3-ospf-1]area 0
[Router3-ospf-1-area-0.0.0.0]network 30.0.0.0 0.255.255.255
```

计算机网络实践教程

```
[Router3-ospf-1-area-0.0.0.0]network 3.3.3.3 0.0.0.0
```

在 Router1 上使用 ping 命令测试 Router1 的环回接口与 Router3 的环回接口之间的连通性。

```
<Router1>ping -a 1.1.1.1 3.3.3.3
 ping 3.3.3.3: 56  data bytes, press CTRL_C to break
   Reply from 3.3.3.3: bytes=56 Sequence=1 ttl=254 time=20 ms
   Reply from 3.3.3.3: bytes=56 Sequence=2 ttl=254 time=20 ms
   Reply from 3.3.3.3: bytes=56 Sequence=3 ttl=254 time=30 ms
   Reply from 3.3.3.3: bytes=56 Sequence=4 ttl=254 time=20 ms
   Reply from 3.3.3.3: bytes=56 Sequence=5 ttl=254 time=20 ms
 --- 3.3.3.3 ping statistics ---
   5 packet(s) transmitted
   5 packet(s) received
   0.00% packet loss
   round-trip min/avg/max = 20/22/30 ms
```

结果显示，通信正常，其他路由器之间使用同样的方法进行测试。

3. 配置基本 ACL

在路由器 Router2 上使用 acl 命令创建一个编号为 2022 的 ACL。

```
[Router2]acl 2022
[Router2-acl-basic-2022]
```

根据需求，在 Router2 上使用 rule 命令配置 ACL 规则，指定规则 ID 为 5，拒绝所有来自 1.1.1.1 的数据穿过 Router2。

```
[Router2-acl-basic-2022]rule 5 deny source 1.1.1.1 0.0.0.0
```

ACL 配置完成后，使用 traffic-filter 命令和 inbound 命令，在 Router2 靠近 1.1.1.1 这个网络接口（GE 0/0/0）的入站方向进行配置。

```
[Router2]interface GigabitEthernet 0/0/0
[Router2-GigabitEthernet0/0/0]traffic-filter inbound acl 2022
```

4. 基本 ACL 测试

在 Router1 上，使用 ping 命令，以 Router1 的环回地址 1.1.1.1 为源地址，测试与 Router3 的环回地址的连通性。

```
<Router1>ping -a 1.1.1.1 3.3.3.3
 ping 3.3.3.3: 56  data bytes, press CTRL_C to break
   Request time out
   Request time out
   Request time out
   Request time out
   Request time out
 --- 3.3.3.3 ping statistics ---
   5 packet(s) transmitted
   0 packet(s) received
   100.00% packet loss
```

结果显示，原本可以 ping 通的测试现在变成超时，不能通信。这说明 ACL 起作用了。如果不限定源地址为 1.1.1.1，直接让 Router1 以接口 GE 0/0/0 去测试与 Router3 的连通性。在 Router1 上，直接使用 ping 命令测试 Router1 与 Router3 的环回地址的连通性。

```
<Router1>ping 3.3.3.3
 ping 3.3.3.3: 56  data bytes, press CTRL_C to break
   Reply from 3.3.3.3: bytes=56 Sequence=1 ttl=254 time=30 ms
   Reply from 3.3.3.3: bytes=56 Sequence=2 ttl=254 time=20 ms
   Reply from 3.3.3.3: bytes=56 Sequence=3 ttl=254 time=30 ms
   Reply from 3.3.3.3: bytes=56 Sequence=4 ttl=254 time=30 ms
   Reply from 3.3.3.3: bytes=56 Sequence=5 ttl=254 time=20 ms
 --- 3.3.3.3 ping statistics ---
   5 packet(s) transmitted
   5 packet(s) received
   0.00% packet loss
   round-trip min/avg/max = 20/26/30 ms
```

结果显示，从 Router1 直接测试连通性，双方可以通信。表明只有以 1.1.1.1 为源地址的通信会被拒绝，测试成功。

8.2 高级 ACL 配置

一、原理概述

高级访问控制列表可以使用数据包的源地址、目的地址、IP 承载的协议类型、针对协议的特性，如 TCP 的源端口、目的端口及 ICMP 协议的类型、代码等内容的定义规则。可以利用高级访问控制列表定义比基本访问控制列表更准确、更丰富、更灵活的规则。

高级访问控制列表的实现，同样可以分为三个步骤：访问控制列表的创建、基本访问控制列表的规则制定、将具体的 ACL 应用到相应接口上。

二、实验目的

1. 掌握配置高级 ACL 的方法
2. 理解基本 ACL 与高级 ACL 的区别

三、实验内容

本实验模拟一个企业网络环境。Rourter3 为企业总部核心路由器，Router1 为分公司边界路由器。分公司通过远程方式管理核心路由器 Router3，要求 Router1 所连接的网络主机可以访问 Router3，其他设备均不能访问。同时要求只能管理 Router3 上的 3.3.3.3 这台服务器，另一台同样直连 Router3 的服务器 4.4.4.4 不能被管理（本实验内网主机使用环回接口模拟）。

四、实验拓扑

高级 ACL 配置拓扑如图 8.2 所示。

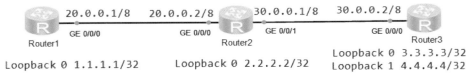

图 8.2　配置高级 ACL 的拓扑

五、实验地址分配

本实验地址分配如表 8.2 所示。

表 8.2　实验地址分配

设备	接口	IP 地址	子网掩码	默认网关
Router1（AR2220）	GE 0/0/0	20.0.0.1	255.0.0.0	N/A
	Loopback 0	1.1.1.1	255.255.255.255	N/A
Router2（AR2220）	GE 0/0/0	20.0.0.2	255.0.0.0	N/A
	GE 0/0/1	30.0.0.1	255.0.0.0	N/A
	Loopback 0	2.2.2.2	255.255.255.255	N/A
Router3（AR2220）	GE 0/0/0	30.0.0.2	255.0.0.0	N/A
	Loopback 0	3.3.3.3	255.255.255.255	N/A
	Loopback 1	4.4.4.4	255.255.255.255	N/A

六、实验步骤

1. 基本配置

根据实验地址分配表进行相应的基本配置,并使用 ping 命令测试各直连线的连通性。

1) Router1 的基本配置

```
<Huawei>system-view
[Huawei]sysname Router1
[Router1]interface GigabitEthernet 0/0/0
[Router1-GigabitEthernet0/0/0]ip address 20.0.0.1 8
[Router1-GigabitEthernet0/0/0]interface loopback 0
[Router1-LoopBack0]ip address 1.1.1.1 32
```

2) Router2 的基本配置

```
<Huawei>system-view
[Huawei]sysname Router2
[Router2]interface GigabitEthernet 0/0/0
[Router2-GigabitEthernet0/0/0]ip address 20.0.0.2 8
[Router2-GigabitEthernet0/0/0]interface GigabitEthernet 0/0/1
[Router2-GigabitEthernet0/0/1]ip address 30.0.0.1 8
[Router2-GigabitEthernet0/0/1]interface loopback 0
[Router2-LoopBack0]ip address 2.2.2.2 32
```

3) Router3 的基本配置

```
<Huawei>system-view
[Huawei]sysname Router3
[Router3]interface GigabitEthernet 0/0/0
[Router3-GigabitEthernet0/0/0]ip address 30.0.0.2 8
[Router3-GigabitEthernet0/0/0]interface loopback 0
[Router3-LoopBack0]ip address 3.3.3.3 32
[Router3-LoopBack0]interface loopback 1
[Router3-LoopBack1]ip address 4.4.4.4 32
```

2. OSPF 配置

路由器基本配置完成后,使用 OSPF 配置动态路由表。

```
[Router1]ospf 1
[Router1-ospf-1]area 0
```

计算机网络实践教程

```
[Router1-ospf-1-area-0.0.0.0]network 20.0.0.0 0.255.255.255
[Router1-ospf-1-area-0.0.0.0]network 1.1.1.1 0.0.0.0
---------------------------------------------------------
[Router2]ospf 1
[Router2-ospf-1]area 0
[Router2-ospf-1-area-0.0.0.0]network 20.0.0.0 0.255.255.255
[Router2-ospf-1-area-0.0.0.0]network 30.0.0.0 0.255.255.255
[Router2-ospf-1-area-0.0.0.0]network 2.2.2.2 0.0.0.0
---------------------------------------------------------
[Router3]ospf 1
[Router3-ospf-1]area 0
[Router3-ospf-1-area-0.0.0.0]network 30.0.0.0 0.255.255.255
[Router3-ospf-1-area-0.0.0.0]network 3.3.3.3 0.0.0.0
[Router3-ospf-1-area-0.0.0.0]network 4.4.4.4 0.0.0.0
```

配置完成后，测试 Router1 的环回接口与 Router3 的环回接口之间的连通性。

```
<Router1>ping -a 1.1.1.1 3.3.3.3
 ping 3.3.3.3: 56  data bytes, press CTRL_C to break
  Reply from 3.3.3.3: bytes=56 Sequence=1 ttl=254 time=20 ms
  Reply from 3.3.3.3: bytes=56 Sequence=2 ttl=254 time=20 ms
  Reply from 3.3.3.3: bytes=56 Sequence=3 ttl=254 time=30 ms
  Reply from 3.3.3.3: bytes=56 Sequence=4 ttl=254 time=20 ms
  Reply from 3.3.3.3: bytes=56 Sequence=5 ttl=254 time=20 ms
 --- 3.3.3.3 ping statistics ---
  5 packet(s) transmitted
  5 packet(s) received
  0.00% packet loss
  round-trip min/avg/max = 20/22/30 ms
```

结果显示，通信正常，其他路由器之间使用同样的方法进行测试。

3. 配置 Telnet

在 Router3 上做 Telnet 相关配置，配置用户名为 hbeutc。

```
[Router2]user-interface vty 0 4
[Router2-ui-vty0-4]authentication-mode password
Please configure the login password (maximum length 16):hbeutc
```

配置完后，测试在 Router1 上建立与 Router3 的环回接口 0 的 IP 地址的 Telnet 连接。

```
<Router1>telnet -a 1.1.1.1 3.3.3.3
  Press CTRL_] to quit telnet mode
```

第 8 章 访问控制列表 ACL 配置

```
  Trying 3.3.3.3 ...
  Connected to 3.3.3.3 ...
Login authentication
Password:          //输入密码hbeutc，但系统不显示
<Router3>
```

结果显示，Router1 可以登录 Router3。

再测试在 Router1 上建立与 Router3 的环回接口 1 的 IP 地址的 Telnet 连接。

```
<Router1>telnet -a 1.1.1.1 4.4.4.4
  Press CTRL_] to quit telnet mode
  Trying 4.4.4.4 ...
  Connected to 4.4.4.4 ...
Login authentication
Password:          //输入密码hbeutc，但系统不显示
<Router3>
```

结果显示，只要是路由可达的设备，并拥有 Telnet 密码，都可以成功登录。

4．配置高级 ACL

根据需求分析，Router1 的环回接口只能通过 Router3 上的 3.3.3.3 进行 Telnet 访问，但是不能通过 4.4.4.4 访问。如果要 Router1 只能通过访问 Router3 的环回接口 0 地址登录设备，即同时匹配数据包的源地址和目的地址实现过滤，此时通过基本 ACL 是无法实现的，因为基本 ACL 只能通过匹配源地址实现过滤，因此要使用高级 ACL。

在 Router3 上使用 acl 命令创建一个高级 ACL 3008。

```
[Router3]acl 3008
```

使用 rule 命令配置 ACL，允许源 IP 地址为 1.1.1.1、目的 IP 地址为 3.3.3.3 的数据包通过。

```
[Router3-acl-adv-3008]rule permit ip source 1.1.1.1 0 destination 3.3.3.3 0
```

配置完成后，使用 display acl all 命令查看 ACL 配置信息。

```
[Router3]display acl all
Total quantity of nonempty ACL number is 1
Advanced ACL 3008, 1 rule
Acl's step is 5
 rule 5 permit ip source 1.1.1.1 0 destination 3.3.3.3 0
```

计算机网络实践教程

结果显示，在不指定规则 ID 的情况下，默认步长为 5，第一条规则的 ID 即为 5。

将 ACL 3008 调用在 VTY 下，使用 inbound 命令，即从外向路由器 Router3 的方向上调用。

```
[Router3]user-interface vty 0 4
[Router3-ui-vty0-4]acl 3008 inbound
```

配置完成后，在 Router1 上使用环回地址 1.1.1.1 访问 3.3.3.3。

```
<Router1>telnet -a 1.1.1.1 3.3.3.3
  Press CTRL_] to quit telnet mode
  Trying 3.3.3.3 ...
  Connected to 3.3.3.3 ...
Login authentication
Password:
<Router3>
```

结果显示，可以访问。再在 Router1 上使用环回地址 1.1.1.1 访问 4.4.4.4。

```
<Router1>telnet -a 1.1.1.1 4.4.4.4
  Press CTRL_] to quit telnet mode
  Trying 4.4.4.4 ...
  Error: Can't connect to the remote host
<Router1>
```

结果显示，规则起作用了，Router1 不能使用环回接口地址访问 4.4.4.4。

高级 ACL 还可以实现对源端口、目的端口、协议号等信息的匹配，功能非常强大。

第 9 章
网络地址转换 NAT 配置

9.1 静态 NAT 配置

一、原理概述

当专用网内部的一些主机本来已经分配到了本地 IP 地址（私有 IP 地址），但又想和互联网上的主机通信时，可使用 NAT（network address translation）方法。这种方法需要在专用网连接到互联网的路由器上安装 NAT 软件。装有 NAT 软件的路由器称为 NAT 路由器，它至少有一个有效的外部全球 IP 地址（公有 IP 地址）。这样，所有使用本地地址的主机在和外界通信时，都要在 NAT 路由器上将其本地 IP 地址转换成全球 IP 地址，才能和互联网连接。这种通过使用少量的全球 IP 地址代表较多的私有 IP 地址的方式，将有助于减缓可用的 IP 地址空间的枯竭。

静态转换是指将内部网络的私有 IP 地址转换为公有 IP 地址，IP 地址对是一对一的，是一成不变的，某个私有 IP 地址只转换为某个公有 IP 地址。借助于静态转换，可以实现外部网络对内部网络中某些特定设备（如服务器）的访问。

二、实验目的

掌握静态 NAT 的配置方法

三、实验内容

本实验模拟企业网络环境，Router1 为边界路由器，Router2 模拟 Internet 资源。通过静态 NAT 方法，分别将本地网络 PC1、PC2 和 PC3 的私有 IP 地址映射为 Router1 的公有 IP 地址。

四、实验拓扑

静态 NAT 配置拓扑如图 9.1 所示。

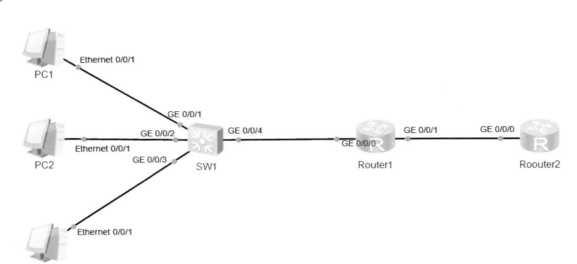

图 9.1 静态 NAT 配置拓扑

五、实验地址分配

本实验 IP 地址分配如表 9.1 所示。

表 9.1 静态 NAT 配置 IP 地址分配表

设备	接口	IP 地址	子网掩码	默认网关
Router1（AR2220）	GE 0/0/0	192.168.1.254	255.255.255.0	N/A
	GE 0/0/1	211.85.8.18	255.255.255.0	N/A
Router2（AR2220）	GE 0/0/0	211.85.8.16	255.255.255.0	N/A
	Loopback 0	211.85.9.16	255.255.255.0	N/A
PC1	Ethernet 0/0/1	192.168.1.1	255.255.255.0	192.168.1.254
PC2	Ethernet 0/0/1	192.168.1.2	255.255.255.0	192.168.1.254
PC3	Ethernet 0/0/1	192.168.1.3	255.255.255.0	192.168.1.254

六、实验步骤

1. PC 的基本配置

PC1 的基本配置如图 9.2 所示。

第 9 章 网络地址转换 NAT 配置

图 9.2 PC1 的基本配置

PC2 的基本配置如图 9.3 所示。

图 9.3 PC2 的基本配置

PC3 的基本配置如图 9.4 所示。

图 9.4 PC3 的基本配置

2. 路由器的基本配置

根据实验 IP 地址分配表，完成各路由器的基本配置。

1) Router1 的基本配置

```
<Huawei>system-view
[Huawei]sysname Router1
[Router1]interface GigabitEthernet 0/0/0
[Router1-GigabitEthernet0/0/0]ip address 192.168.1.254 24
[Router1-GigabitEthernet0/0/0]interface GigabitEthernet 0/0/1
[Router1-GigabitEthernet0/0/1]ip address 211.85.8.18 24
```

163

2) 配置路由器 Router2

```
<Huawei>system-view
[Huawei]sysname Router2
[Router2]interface GigabitEthernet 0/0/0
[Router2-GigabitEthernet0/0/0]ip address 211.85.8.16 24
[Router2-GigabitEthernet0/0/0]interface loopback 0
[Router2-LoopBack0]ip address 211.85.9.16 24
```

3) 测试网络的连通性

基本配置完毕后，依次检测各直连链路的连通性。在 Router1 上使用 **ping** 命令测试与 Router2 之间的连通性。

```
[Router1]ping 211.85.8.16
  ping 211.85.8.16: 56 data bytes, press CTRL_C to break
    Reply from 211.85.8.16: bytes=56 Sequence=1 ttl=255 time=20 ms
    Reply from 211.85.8.16: bytes=56 Sequence=2 ttl=255 time=20 ms
    Reply from 211.85.8.16: bytes=56 Sequence=3 ttl=255 time=10 ms
    Reply from 211.85.8.16: bytes=56 Sequence=4 ttl=255 time=30 ms
    Reply from 211.85.8.16: bytes=56 Sequence=5 ttl=255 time=20 ms
  --- 211.85.8.16 ping statistics ---
    5 packet(s) transmitted
    5 packet(s) received
    0.00% packet loss
    round-trip min/avg/max = 10/20/30 ms
```

结果显示，该直连网段是通的。其他直连网段连通性测试省略。各直连链路连通后再进行后续操作。

在 PC1 上使用 **ping** 命令测试与 Router2 的连通性。

```
PC>ping 211.85.8.16
ping 211.85.8.16: 32 data bytes, Press Ctrl_C to break
Request timeout!
Request timeout!
Request timeout!
Request timeout!
Request timeout!
--- 211.85.8.16 ping statistics ---
  5 packet(s) transmitted
  0 packet(s) received
  100.00% packet loss
```

结果显示,PC1 无法访问 Router2 的 GE 0/0/0 接口。

3 配置静态 NAT

在 Router1 上配置访问外网的默认路由。

```
[Router1]ip route-static 0.0.0.0 0.0.0.0 211.85.8.16
```

由于内网使用的都是私有 IP 地址,主机无法直接访问公网。需要在 Router1 上配置 NAT 地址转换,将私有 IP 地址转换为公有 IP 地址。

在 Router1 的 GE 0/0/1 接口下依次使用 **nat static** 命令将私有 IP 地址 192.168.1.1、192.168.1.2 和 192.168.1.3 分别转换为 211.85.8.20、211.85.8.21 和 211.85.8.22 等公有 IP 地址。

```
[Router1]interface GigabitEthernet 0/0/1
  [Router1-GigabitEthernet0/0/1]nat static global 211.85.8.20 inside 192.168.1.1
  [Router1-GigabitEthernet0/0/1]nat static global 211.85.8.21 inside 192.168.1.2
  [Router1-GigabitEthernet0/0/1]nat static global 211.85.8.22 inside 192.168.1.3
```

配置完成后,在 Router1 上使用 **display nat static** 命令查看 NAT 静态配置信息。

```
<Router1>display nat static
 Static Nat Information:
 Interface  : GigabitEthernet0/0/1
   Global IP/Port     : 211.85.8.20/----
   Inside IP/Port     : 192.168.1.1/----
   Protocol : ----
   VPN instance-name  : ----
   Acl number         : ----
   Netmask : 255.255.255.255
   Description : ----
   Global IP/Port     : 211.85.8.21/----
   Inside IP/Port     : 192.168.1.2/----
   Protocol : ----
   VPN instance-name  : ----
   Acl number         : ----
   Netmask : 255.255.255.255
   Description : ----
   Global IP/Port     : 211.85.8.22/----
   Inside IP/Port     : 192.168.1.3/----
   Protocol : ----
   VPN instance-name  : ----
```

```
    Acl number             : ----
    Netmask : 255.255.255.255
```

4. 配置 NAT Server

配置 NAT Server 并使用公有 IP 地址对外公布服务器地址，然后开启 NAT ALG 功能，因为对于封装在 IP 数据包中的应用层协议报文，正常的 NAT 转换会导致错误，在开启某应用协议的 NAT ALG 功能后，该应用协议报文可以正常进行 NAT 转换；否则，该应用协议不能正常工作。

假定 PC3 为 FTP 服务器。在 Roouter1 上使用 **nat server** 命令将 FTP 服务器对私有地址 192.168.1.3，映射到公有 IP 地址 211.85.8.22，FTP 是基于 TCP 的协议。使用 **nat alg** 命令开启 NAT ALG 功能。

```
[Router1]interface GigabitEthernet 0/0/1
[Router1-GigabitEthernet0/0/1]nat server protocol tcp global 211.85.8.22 ftp inside 192.168.1.3 ftp
[Router1-GigabitEthernet0/0/1]quit
[Router1]nat alg ftp enable
```

配置完成后，Internet 用户可以通过 211.85.8.22 访问内网的 FTP 服务器 192.168.1.3。

5. 测试

在 PC1 上使用 ping 命令测试与外网（Router2 的 GE 0/0/0 接口）的连通性。

```
PC>ping 211.85.8.16
ping 211.85.8.16: 32 data bytes, Press Ctrl_C to break
From 211.85.8.16: bytes=32 seq=1 ttl=254 time=63 ms
From 211.85.8.16: bytes=32 seq=2 ttl=254 time=46 ms
From 211.85.8.16: bytes=32 seq=3 ttl=254 time=63 ms
From 211.85.8.16: bytes=32 seq=4 ttl=254 time=31 ms
From 211.85.8.16: bytes=32 seq=5 ttl=254 time=31 ms
--- 211.85.8.16 ping statistics ---
  5 packet(s) transmitted
  5 packet(s) received
  0.00% packet loss
  round-trip min/avg/max = 31/46/63 ms
```

在 PC1 上使用 ping 命令测试与外网（Router2 的 Loopback 0 接口）的连通性。

```
PC>ping 211.85.9.16
ping 211.85.9.16: 32 data bytes, Press Ctrl_C to break
```

```
From 211.85.9.16: bytes=32 seq=1 ttl=254 time=62 ms
From 211.85.9.16: bytes=32 seq=2 ttl=254 time=31 ms
From 211.85.9.16: bytes=32 seq=3 ttl=254 time=32 ms
From 211.85.9.16: bytes=32 seq=4 ttl=254 time=15 ms
From 211.85.9.16: bytes=32 seq=5 ttl=254 time=47 ms
--- 211.85.9.16 ping statistics ---
  5 packet(s) transmitted
  5 packet(s) received
  0.00% packet loss
  round-trip min/avg/max = 15/37/62 ms
```

结果显示，PC1 通过静态 NAT 地址转换已经可以成功访问外网。

在路由器 Router1 的 GE 0/0/1 接口上抓包查看 NAT 地址转换是否成功，先设置抓包，再在 PC1 上使用 **ping** 命令测试与 Router2 的 GE 0/0/0 接口的连通性。结果如图 9.5 所示。

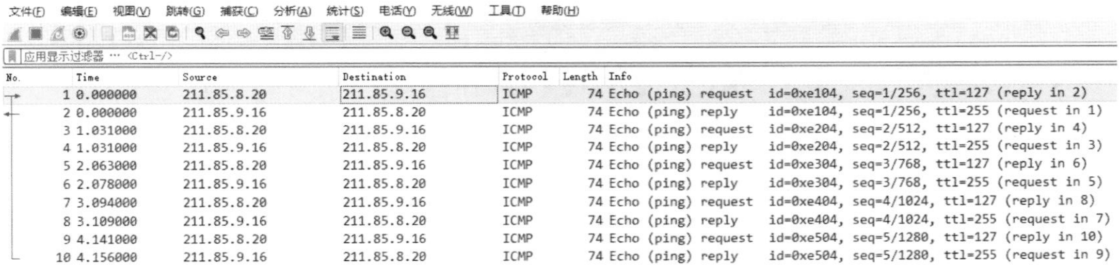

No.	Time	Source	Destination	Protocol	Length	Info
1	0.000000	211.85.8.20	211.85.9.16	ICMP	74	Echo (ping) request id=0xe104, seq=1/256, ttl=127 (reply in 2)
2	0.000000	211.85.9.16	211.85.8.20	ICMP	74	Echo (ping) reply id=0xe104, seq=1/256, ttl=255 (request in 1)
3	1.031000	211.85.8.20	211.85.9.16	ICMP	74	Echo (ping) request id=0xe204, seq=2/512, ttl=127 (reply in 4)
4	1.031000	211.85.9.16	211.85.8.20	ICMP	74	Echo (ping) reply id=0xe204, seq=2/512, ttl=255 (request in 3)
5	2.063000	211.85.8.20	211.85.9.16	ICMP	74	Echo (ping) request id=0xe304, seq=3/768, ttl=127 (reply in 6)
6	2.078000	211.85.9.16	211.85.8.20	ICMP	74	Echo (ping) reply id=0xe304, seq=3/768, ttl=255 (request in 5)
7	3.094000	211.85.8.20	211.85.9.16	ICMP	74	Echo (ping) request id=0xe404, seq=4/1024, ttl=127 (reply in 8)
8	3.109000	211.85.9.16	211.85.8.20	ICMP	74	Echo (ping) reply id=0xe404, seq=4/1024, ttl=255 (request in 7)
9	4.141000	211.85.8.20	211.85.9.16	ICMP	74	Echo (ping) request id=0xe504, seq=5/1280, ttl=127 (reply in 10)
10	4.156000	211.85.9.16	211.85.8.20	ICMP	74	Echo (ping) reply id=0xe504, seq=5/1280, ttl=255 (request in 9)

图 9.5　静态 NAT 抓包截图

结果显示，Router1 已经成功把来自 PC1 的 ICMP 报文的源地址 192.168.1.1 转换成公有 IP 地址 211.85.8.20。在 R2 上使用环回口 loopback 0 模拟外网用户访问 PC1(在 Router2 上使用 **ping -a** 命令在多个接口的设备中指明在 211.85.9.16 接口上测试与 PC1 的连通性），并在 PC1 的 Ethernet 0/0/1 接口上抓包观察，如图 9.6 所示。

```
<Router2>ping -a 211.85.9.16 211.85.8.20
  ping 211.85.8.20: 56  data bytes, press CTRL_C to break
    Reply from 211.85.8.20: bytes=56 Sequence=1 ttl=127 time=40 ms
    Reply from 211.85.8.20: bytes=56 Sequence=2 ttl=127 time=50 ms
    Reply from 211.85.8.20: bytes=56 Sequence=3 ttl=127 time=60 ms
    Reply from 211.85.8.20: bytes=56 Sequence=4 ttl=127 time=70 ms
    Reply from 211.85.8.20: bytes=56 Sequence=5 ttl=127 time=60 ms
  --- 211.85.8.20 ping statistics ---
    5 packet(s) transmitted
    5 packet(s) received
    0.00% packet loss
    round-trip min/avg/max = 40/56/70 ms
```

No.	Time	Source	Destination	Protocol	Length	Info
4	4.750000	211.85.9.16	192.168.1.1	ICMP	98	Echo (ping) request id=0xceab, seq=256/1, ttl=254 (reply in 5)
5	4.750000	192.168.1.1	211.85.9.16	ICMP	98	Echo (ping) reply id=0xceab, seq=256/1, ttl=128 (request in 4)
6	5.250000	211.85.9.16	192.168.1.1	ICMP	98	Echo (ping) request id=0xceab, seq=512/2, ttl=254 (reply in 7)
7	5.250000	192.168.1.1	211.85.9.16	ICMP	98	Echo (ping) reply id=0xceab, seq=512/2, ttl=128 (request in 6)
8	5.750000	211.85.9.16	192.168.1.1	ICMP	98	Echo (ping) request id=0xceab, seq=768/3, ttl=254 (reply in 9)
9	5.750000	192.168.1.1	211.85.9.16	ICMP	98	Echo (ping) reply id=0xceab, seq=768/3, ttl=128 (request in 8)
10	6.250000	211.85.9.16	192.168.1.1	ICMP	98	Echo (ping) request id=0xceab, seq=1024/4, ttl=254 (reply in 11)
11	6.266000	192.168.1.1	211.85.9.16	ICMP	98	Echo (ping) reply id=0xceab, seq=1024/4, ttl=128 (request in 10)
12	6.735000	211.85.9.16	192.168.1.1	ICMP	98	Echo (ping) request id=0xceab, seq=1280/5, ttl=254 (reply in 13)
13	6.735000	192.168.1.1	211.85.9.16	ICMP	98	Echo (ping) reply id=0xceab, seq=1280/5, ttl=128 (request in 12)

图 9.6　抓包结果

结果显示，PC1 的私有 IP 地址被转换为唯一的公有 IP 地址，外网用户也能主动访问 PC1，且数据包在经过 Router1 进入内网的时候，Router1 把目的 IP 地址转换为与公有地址 211.85.8.20 对应的私有 IP 地址 192.168.1.1 发给 PC1。

9.2　动态 NAT 配置

一、原理概述

动态转换是指将内部网络的私有 IP 地址转换为公用 IP 地址时，IP 地址是不确定的，是随机的，所有被授权访问 Internet 的私有 IP 地址可随机转换为任何指定的合法 IP 地址。也就是说，只要指定哪些内部地址可以进行转换，以及用哪些合法地址作为外部地址时，就可以进行动态转换。动态转换可以使用多个合法外部地址集。当 ISP 提供的合法 IP 地址略少于网络内部的计算机数量时，可以采用动态转换的方式。

二、实验目的

掌握动态 NAT 的配置方法

三、实验内容

本实验模拟企业网络环境，Router1 为边界路由器，Router2 模拟 Internet 资源。通过动态 NAT 方法，分别将本地网络 PC1、PC2 和 PC3 的私有 IP 地址映射为 Router1 的公有 IP 地址。

四、实验拓扑

动态 NAT 配置拓扑如图 9.7 所示。

第 9 章 网络地址转换 NAT 配置

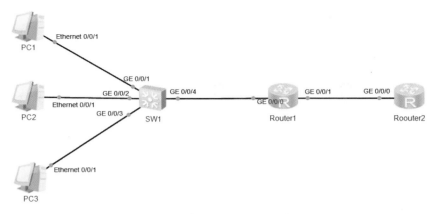

图 9.7 静态 NAT 配置拓扑

五、实验地址分配

本实验 IP 地址分配如表 9.2 所示。

表 9.2 静态 NAT 配置 IP 地址分配表

设备	接口	IP 地址	子网掩码	默认网关
Router1（AR2220）	GE 0/0/0	192.168.1.254	255.255.255.0	N/A
	GE 0/0/1	211.85.8.18	255.255.255.0	N/A
Router2（AR2220）	GE 0/0/0	211.85.8.16	255.255.255.0	N/A
	Loopback 0	211.85.9.16	255.255.255.0	N/A
PC1	Ethernet 0/0/1	192.168.1.1	255.255.255.0	192.168.1.254
PC2	Ethernet 0/0/1	192.168.1.2	255.255.255.0	192.168.1.254
PC3	Ethernet 0/0/1	192.168.1.3	255.255.255.0	192.168.1.254

六、实验步骤

1. PC 的基本配置

PC1 的基本配置如图 9.8 所示。

图 9.8 PC1 的基本配置

PC2 的基本配置如图 9.9 所示。

图 9.9　PC2 的基本配置

PC3 的基本配置如图 9.10 所示。

图 9.10　PC3 的基本配置

2. 路由器的基本配置

根据实验地址分配表完成 Router1 和 Router2 的基本配置。

1) Router1 的基本配置

```
<Huawei>system-view
[Huawei]sysname Router1
[Router1]interface GigabitEthernet 0/0/0
[Router1-GigabitEthernet0/0/0]ip address 192.168.1.254 24
[Router1-GigabitEthernet0/0/0]interface GigabitEthernet 0/0/1
[Router1-GigabitEthernet0/0/1]ip address 211.85.8.18 24
```

2) 配置路由器 Router2

```
<Huawei>system-view
[Huawei]sysname Router2
[Router2]interface GigabitEthernet 0/0/0
[Router2-GigabitEthernet0/0/0]ip address 211.85.8.16 24
[Router2-GigabitEthernet0/0/0]interface loopback 0
[Router2-LoopBack0]ip address 211.85.9.16 24
```

3) 测试网络的连通性

基本配置完毕后,依次检测各直连链路的连通性。在 Router1 上使用 ping 命令测试与 Router2 之间的连通性。

```
[Router1]ping 211.85.8.16
 ping 211.85.8.16: 56 data bytes, press CTRL_C to break
   Reply from 211.85.8.16: bytes=56 Sequence=1 ttl=255 time=20 ms
   Reply from 211.85.8.16: bytes=56 Sequence=2 ttl=255 time=20 ms
   Reply from 211.85.8.16: bytes=56 Sequence=3 ttl=255 time=10 ms
   Reply from 211.85.8.16: bytes=56 Sequence=4 ttl=255 time=30 ms
   Reply from 211.85.8.16: bytes=56 Sequence=5 ttl=255 time=20 ms
 --- 211.85.8.16 ping statistics ---
   5 packet(s) transmitted
   5 packet(s) received
   0.00% packet loss
   round-trip min/avg/max = 10/20/30 ms
```

结果显示,该直连网段是通的。其他直连网段连通性测试省略。各直连链路连通后再进行后续操作。

在 PC1 上使用 ping 命令测试与 Router2 的连通性。

```
PC>ping 211.85.8.16
ping 211.85.8.16: 32 data bytes, Press Ctrl_C to break
Request timeout!
Request timeout!
Request timeout!
Request timeout!
Request timeout!
--- 211.85.8.16 ping statistics ---
   5 packet(s) transmitted
   0 packet(s) received
   100.00% packet loss
```

结果显示,PC1 无法访问 Router2 的 GE 0/0/0 接口。

3. 配置动态 NAT

在 Router1 上配置访问外网的默认路由。

```
[Router1]ip route-static 0.0.0.0 0.0.0.0 211.85.8.16
```

由于内网使用的都是私有 IP 地址 192.168.1.0/24 网段,企业可用公网地址比较多,地址范围为 211.85.8.20~211.85.8.50。需要在 Router1 上配置 NAT 地址转换，将私有网段动态映射到公网 IP 地址池。

在 Router1 上使用 **nat address-group** 命令配置 NAT 地址池，地址池索引号为 1，设置起始和终止地址分别为 211.85.8.20 和 211.85.8.50。

```
[Router1]nat address-group 1 211.85.8.20 211.85.8.50
```

创建基本 ACL 2022，匹配 192.168.1.0/24 地址段。

```
[Router1]acl 2022        //编号为2000~2999的ACL为基本ACL
[Router1-acl-basic-2022]rule 5 permit source 192.168.1.0 0.0.0.255
```

在 Router1 的 GE 0/0/1 接口下使用 **nat outbound** 命令将 ACL 2022 与地址池相关联，使得 ACL 中规定的地址可以使用地址池进行地址转换，no-pat 表示只做 IP 地址转换，不做端口转换。

```
[Router1]interface GigabitEthernet 0/0/1
[Router1-GigabitEthernet0/0/1]nat outbound 2022 address-group 1 no-pat
```

配置完成后，在 Router1 上使用 **display nat outbound** 命令查看 NAT Outbound 信息。

```
[Router1]display nat outbound
NAT Outbound Information:
--------------------------------------------------------------------
  Interface              Acl    Address-group/IP/Interface    Type
--------------------------------------------------------------------
  GigabitEthernet0/0/1   2022             1                   no-pat
--------------------------------------------------------------------
  Total : 1
```

在 PC2 上使用 **ping** 命令测试与外网的连通性，并在 Router1 的接口 GE 0/0/1 上抓包，观察地址转换情况，如图 9.11 所示。

```
PC>ping 211.85.9.16
ping 211.85.9.16: 32 data bytes, Press Ctrl_C to break
From 211.85.9.16: bytes=32 seq=1 ttl=254 time=62 ms
From 211.85.9.16: bytes=32 seq=2 ttl=254 time=47 ms
From 211.85.9.16: bytes=32 seq=3 ttl=254 time=47 ms
From 211.85.9.16: bytes=32 seq=4 ttl=254 time=63 ms
From 211.85.9.16: bytes=32 seq=5 ttl=254 time=31 ms
```

第 9 章　网络地址转换 NAT 配置

```
--- 211.85.9.16 ping statistics ---
 5 packet(s) transmitted
 5 packet(s) received
 0.00% packet loss
 round-trip min/avg/max = 31/50/63 ms
```

结果显示，PC2 与 Router2 通信正常。

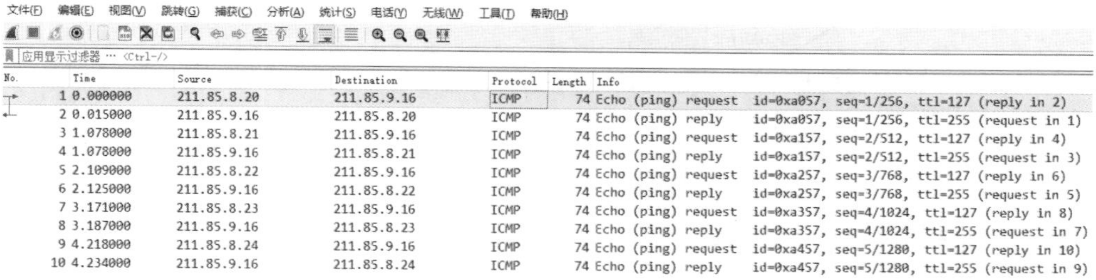

图 9.11　抓包结果

结果显示 PC2 可以访问外网，通过抓包分析，来自 PC1 的 ICMP 数据包在 Router1 的 GE 0/0/1 接口上源地址 192.168.1.2 被替换为地址池中的第一个 IP 地址 211.85.8.20。

9.3　NAPT 配置

一、原理概述

网络地址端口转换（network address port translation，NAPT）是指改变外出数据包的源端口并进行端口转换。采用 NAPT 方式，内部网络的所有主机均可共享一个合法外部 IP 地址实现对 Internet 的访问，从而可以最大限度地节约 IP 地址资源。同时，又可隐藏网络内部的所有主机，有效避免来自 Internet 的攻击。因此，网络中应用最多的就是 NAPT 方式。

二、实验目的

掌握 NAPT 的配置方法

三、实验内容

本实验模拟企业网络环境，Router1 为边界路由器，Router2 模拟 Internet 资源。通过 NAPT 方法，分别将本地网络 PC1、PC2 和 PC3 的私有 IP 地址映射为 Router1 的唯一公有 IP 地址。

四、实验拓扑

NAPT 配置拓扑如图 9.12 所示。

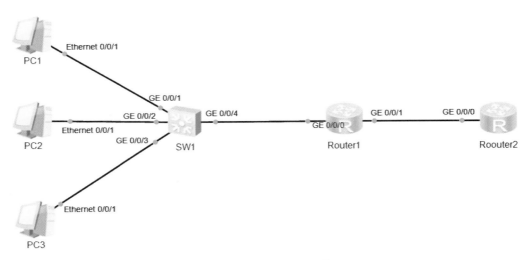

图 9.12　NAPT 配置拓扑

五、实验地址分配

本实验 IP 地址分配如表 9.3 所示。

表 9.3　静态 NAT 配置 IP 地址分配表

设备	接口	IP 地址	子网掩码	默认网关
Router1 （AR2220）	GE 0/0/0	192.168.1.254	255.255.255.0	N/A
	GE 0/0/1	211.85.8.18	255.255.255.0	N/A
Router2 （AR2220）	GE 0/0/0	211.85.8.16	255.255.255.0	N/A
	Loopback 0	211.85.9.16	255.255.255.0	N/A
PC1	Ethernet 0/0/1	192.168.1.1	255.255.255.0	192.168.1.254
PC2	Ethernet 0/0/1	192.168.1.2	255.255.255.0	192.168.1.254
PC3	Ethernet 0/0/1	192.168.1.3	255.255.255.0	192.168.1.254

六、实验步骤

1. PC 的基本配置

PC1 的基本配置如图 9.13 所示。

第 9 章　网络地址转换 NAT 配置

图 9.13　PC1 的基本配置

PC2 的基本配置如图 9.14 所示。

图 9.14　PC2 的基本配置

PC3 的基本配置如图 9.15 所示。

图 9.15　PC3 的基本配置

2. 路由器的基本配置

根据实验地址分配表完成 Router1 和 Router2 的基本配置。

1) Router1 的基本配置

```
<Huawei>system-view
[Huawei]sysname Router1
[Router1]interface GigabitEthernet 0/0/0
[Router1-GigabitEthernet0/0/0]ip address 192.168.1.254 24
[Router1-GigabitEthernet0/0/0]interface GigabitEthernet 0/0/1
[Router1-GigabitEthernet0/0/1]ip address 211.85.8.18 24
```

2) 配置路由器 Router2

```
<Huawei>system-view
[Huawei]sysname Router2
[Router2]interface GigabitEthernet 0/0/0
[Router2-GigabitEthernet0/0/0]ip address 211.85.8.16 24
[Router2-GigabitEthernet0/0/0]interface loopback 0
[Router2-LoopBack0]ip address 211.85.9.16 24
```

3) 测试网络的连通性

基本配置完毕后，依次检测各直连链路的连通性。在 Router1 上使用 ping 命令测试与 Router2 之间的连通性。

```
[Router1]ping 211.85.8.16
  ping 211.85.8.16: 56  data bytes, press CTRL_C to break
    Reply from 211.85.8.16: bytes=56 Sequence=1 ttl=255 time=20 ms
    Reply from 211.85.8.16: bytes=56 Sequence=2 ttl=255 time=20 ms
    Reply from 211.85.8.16: bytes=56 Sequence=3 ttl=255 time=10 ms
    Reply from 211.85.8.16: bytes=56 Sequence=4 ttl=255 time=30 ms
    Reply from 211.85.8.16: bytes=56 Sequence=5 ttl=255 time=20 ms
  --- 211.85.8.16 ping statistics ---
    5 packet(s) transmitted
    5 packet(s) received
    0.00% packet loss
    round-trip min/avg/max = 10/20/30 ms
```

结果显示，该直连网段是通的。其他直连网段连通性测试省略。各直连链路连通后再进行后续操作。

在 PC1 上使用 ping 命令测试与 Router2 的连通性。

```
PC>ping 211.85.8.16
ping 211.85.8.16: 32 data bytes, Press Ctrl_C to break
Request timeout!
Request timeout!
Request timeout!
Request timeout!
Request timeout!
--- 211.85.8.16 ping statistics ---
  5 packet(s) transmitted
  0 packet(s) received
  100.00% packet loss
```

第 9 章 网络地址转换 NAT 配置

结果显示，PC1 无法访问 Router2 的 GE 0/0/0 接口。

3. NAPT 的配置

在 Router1 上配置访问外网的默认路由。

```
[Router1]ip route-static 0.0.0.0 0.0.0.0 211.85.8.16
```

由于内网使用的都是私有 IP 地址 192.168.1.0/24 网段，唯一的公有 IP 地址配置在 Router1 的 GE 0/0/1 接口。NAPT 可以直接借用路由器外网接口 IP 地址为公有 IP 地址，将不同的内网私有 IP 地址映射到同一个公有 IP 地址的不同端口号上，实现了多对一地址转换。

创建基本 ACL 2023，匹配 192.168.1.0/24 地址段。

```
[Router1]acl 2023    //编号为2000~2999的ACL为基本ACL
[Router1-acl-basic-2022]rule 5 permit source 192.168.1.0 0.0.0.255
```

在 Router1 的 GE 0/0/1 接口下使用 **nat outbound** 命令配置 Easy-IP 特性，直接使用接口 IP 地址作为 NAT 转换后的地址。

```
[Router1]interface GigabitEthernet 0/0/1
[Router1-GigabitEthernet0/0/1]nat outbound 2023
```

配置完成后，在 Router1 上使用 **display nat outbound** 命令查看 nat outbound 信息。

```
[Router1]display nat outbound
 NAT Outbound Information:
--------------------------------------------------------------------
   Interface              Acl  Address-group/IP/Interface     Type
--------------------------------------------------------------------
   GigabitEthernet0/0/1   2023           211.85.8.18          easyip
--------------------------------------------------------------------
  Total : 1
```

在 PC1 上使用 UDP 发包工具发送 UDP 数据包到公网地址 211.85.9.16，配置好目的 IP 地址和 UDP 源、目的端口号后，输入字符串后单击"发送"按钮，如图 9.16 所示。

图 9.16　PC1 上发送 UDP 数据包

在 PC2 上使用 UDP 发包工具发送 UDP 数据包到公网地址 211.85.9.16，配置好目的 IP 地址和 UDP 源、目的端口号后，输入字符串后单击"发送"按钮，如图 9.17 所示。

图 9.17　PC2 上发送 UDP 数据包

在 PC3 上使用 UDP 发包工具发送 UDP 数据包到公网地址 211.85.9.16，配置好目的 IP 地址和 UDP 源、目的端口号后，输入字符串后单击"发送"按钮，如图 9.18 所示。

图 9.18　PC3 上发送 UDP 数据包

在 Router1 上使用 **display nat seession protocol udp verbose** 命令查看 NAPT 映射结果。

```
<Router1>display nat session protocol udp verbose
 NAT Session Table Information:
Protocol           : UDP(17)
   SrcAddr  Port Vpn : 192.168.1.1     2560
   DestAddr Port Vpn : 211.85.9.16     2560
   Time To Live      : 120 s
   NAT-Info
     New SrcAddr     : 211.85.8.18
     New SrcPort     : 10250
     New DestAddr    : ----
     New DestPort    : ----
```

```
    Protocol            : UDP(17)
    SrcAddr  Port Vpn   : 192.168.1.2    2560
    DestAddr Port Vpn   : 211.85.9.16    2560
    Time To Live        : 120 s
    NAT-Info
      New SrcAddr       : 211.85.8.18
      New SrcPort       : 10251
      New DestAddr      : ----
      New DestPort      : ----

    Protocol            : UDP(17)
    SrcAddr  Port Vpn   : 192.168.1.3    2560
    DestAddr Port Vpn   : 211.85.9.16    2560
    Time To Live        : 120 s
    NAT-Info
      New SrcAddr       : 211.85.8.18
      New SrcPort       : 10252
      New DestAddr      : ----
      New DestPort      : ----
 Total : 3
```

结果显示，源地址为 192.168.1.1~192.168.1.3 的 UDP 数据包被新源地址 211.85.8.18 替换，而端口号分别被新源端口号 10250、10251 和 10252 替换。Router1 借用 GE 0/0/1 接口公有 IP 地址做 NAT 转换，使用不同的端口号区分不同内网私有 IP 地址数据。以上配置被称为 Easy-IP 方式，此方式是 NAPT 的一种方式，且不需要创建地址池。

NAPT 配置也可以按照动态 NAT 配置的地址池方式来配置，只是将地址池配置命令改为：

```
[Router1]nat address-group 1 211.85.8.18 211.85.8.18
```

其他配置参照动态 NAT 进行配置。

第 10 章 应用服务器配置

10.1 FTP 服务器配置

一、原理概述

文件传输协议（file transfer protocol，FTP）是 Internet 上最早使用的协议之一，在 TCP/IP 协议栈中属于应用层协议，是 Internet 的文件传输标准。主要功能是向用户提供本地和远程主机之间的文件传输。用户可以通过 FTP 客户端程序登录到服务器，进行文件的上传、下载等操作。

二、实验目的

1. 理解 FTP 的应用场景
2. 掌握操作 FTP 服务器的常见命令
3. 掌握保存文件到 FTP 服务器的方法
4. 掌握获取 FTP 服务器文件到本地的方法

三、实验内容

本实验模拟企业网络环境，客户机 Client 为 FTP 客户端设备，需要访问 FTP Server，从服务器上下载或上传文件。出于安全的考虑，为防止服务器病毒文件的感染，不允许用户直接上传文件到 FTP Server。网络管理员在 Router 上设置了限制，使员工不能上传文件到 FTP Server，但是可以从 FTP Server 下载文件，Router 也需要作为客户端从 FTP Server 下载更新文件，同时配置 Router 作为 FTP Server，员工可以上传文件到 Router 上，经过管理员的检测后由 Router 上传到 FTP Server。

第 10 章 应用服务器配置

四、实验拓扑

FTP 服务器配置拓扑如图 10.1 所示。

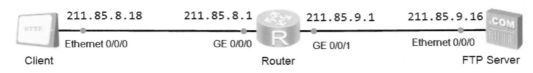

图 10.1　FTP 服务器配置拓扑

五、实验地址分配

实验地址分配如表 10.1 所示。

表 10.1　实验地址分配

设备	接口	IP 地址	子网掩码	默认网关
Router（AR2220）	GE 0/0/0	211.85.8.1	255.255.255.0	N/A
	GE 0/0/1	211.85.9.1	255.255.255.0	N/A
Client	Ethernet 0/0/0	211.85.8.18	255.255.255.0	211.85.8.1
FTP Server	Ethernet 0/0/0	211.85.9.16	255.255.255.0	211.85.9.1

六、实验步骤

1. 基本配置

根据实验地址分配表进行相应的基本配置。

1) Client 的配置

Client 的基本配置如图 10.2 所示。

图 10.2　Client 的基本配置

2) 路由器 Router 的基本配置

```
<Huawei>system-view
[Huawei]sysname Router
```

```
[Router]interface GigabitEthernet 0/0/0
[Router-GigabitEthernet0/0/0]ip address 211.85.8.1 24
[Router-GigabitEthernet0/0/0]interface GigabitEthernet 0/0/1
[Router-GigabitEthernet0/0/1]ip address 211.85.9.1 24
```

3) FTP Server 的基本配置

FTP Server 的基本配置如图 10.3 所示。

图 10.3　FTP　Server 的基本配置

4) 测试

在 Client 客户机上使用 ping 测试功能测试与 FTP Server 的连通性，在目的 IP 地址中填写 211.85.9.16，次数填写 4，单击"发送"按钮，如图 10.4 所示。

图 10.4　Client 与 FTP　Server 连通性测试

结果显示，Client 客户机与 FTP Server 正常通信。

2. 配置 FTP 服务器

1) 配置 FTP Server 服务器

首先，在本地计算机上创建一个文件夹"FTP"，作为 FTP 服务器的文件夹。在该文件夹下创建一个文件夹 ABC 和一个文本文件 TEST.txt。创建完成后，设置 FTP Server 服务器，打开 FTP Server 服务器的配置界面，如图 10.5 所示。选择"服务器信息"标签，选中"FtpServer"单选按钮，右侧的操作栏选择刚刚创建的 FTP 文件夹，单击"启动"按钮。

2) 测试

在 Router 上使用 **ftp** 命令连接 FTP 服务器，登录时默认需要输入用户名和密码，由于服务器上没有设置用户名和密码，每次在 Router 上输入时等同于创建该用户名和密码，本次使用用户名为 admin，密码为 hbeutc。

第 10 章　应用服务器配置

图 10.5　配置 FTP　Server 服务器

```
<Router>ftp 211.85.9.16
Trying 211.85.9.16 ...
Press CTRL+K to abort
Connected to 211.85.9.16.
220 FtpServerTry FtpD for free
User(211.85.9.16:(none)):admin
331 Password required for 211.85.9.16 .
Enter password:                    //输入密码：hbeutc
230 User 211.85.9.16 logged in , proceed
[Router-ftp]
```

结果显示，路由器进入 FTP 配置视图。

在 Router 上使用 ls 命令查看 FTP 服务器的文件夹信息。

```
[Router-ftp]ls
200 Port command okay.
150 Opening ASCII NO-PRINT mode data connection for ls -l.
ABC          //子文件夹
TEST.txt     //文件
226 Transfer finished successfully. Data connection closed.
FTP: 15 byte(s) received in 0.140 second(s) 107.14byte(s)/sec.
[Router-ftp]
```

结果显示，当前文件夹下有一个子文件夹 ABC 和一个文件 TEST.txt。

在 Router 上使用 dir 命令查看详细的文件属性。

```
[Router-ftp]dir
200 Port command okay.
```

183

```
150 Opening ASCII NO-PRINT mode data connection for ls -l.
drwxrwxrwx   1              nogroup              0 Sep 15   2022 ABC
-rwxrwxrwx   1              nogroup              0 Sep 15   2022 TEST.txt
226 Transfer finished successfully. Data connection closed.
FTP: 131 byte(s) received in 0.140 second(s) 935.71byte(s)/sec.
[Router-ftp]
```

结果显示,当前文件夹为 ABC。

在 Router 上使用 get 命令下载 TEST.txt 到本地路由器。

```
[Router-ftp]get test.txt     //文件名不区分大小写
200 Port command okay.
150 Sending test.txt (0 bytes). Mode STREAM Type BINARY
226 Transfer finished successfully. Data connection closed.
FTP: 0 byte(s) received in 0.150 second(s) 0.00byte(s)/sec.
[Router-ftp]
```

结果显示,下载文件成功。

在 Router 上使用 cd 命令进入 ABC 子文件夹。

```
[Router-ftp]cd ABC
250 "/ABC" is current directory.
```

结果显示,当前文件夹为 ABC。

使用 put 命令上传 test.txt 到 FTP 服务器,并命名为 test1.txt。

```
[Router-ftp]put test.txt test1.txt
200 Port command okay.
150 Opening BINARY data connection for test1.txt
226 Transfer finished successfully. Data connection closed.
FTP: 0 byte(s) sent in 0.150 second(s) 0.00byte(s)/sec.
[Router-ftp]
```

结果显示,上传成功。

使用 cd..命令切换到上一级文件夹,使用 ls 命令来查看上传后的结果。

```
[Router-ftp]cd ..                    //当前文件夹为abc
250 "/" is current directory.        //切换到上一级文件夹,客户端默认切换至FTP根目录
[Router-ftp]ls
```

```
200 Port command okay.
150 Opening ASCII NO-PRINT mode data connection for ls -l.
 100%
ABC
TEST.txt
test1.txt
226 Transfer finished successfully. Data connection closed.
FTP: 26 byte(s) received in 0.030 second(s) 866.66byte(s)/sec.
```

3. 配置路由器为 FTP 服务器

在路由器 Router 上使用 **ftp server enable** 命令配置 FTP 服务器。设置 FTP 登录的用户名为 ftp，密码为 hbeutc，配置 FTP 用户访问的目录为"flash："，用户优先级为 15，服务类型为 ftp。

```
[Router]ftp server enable
Info: Succeeded in starting the FTP server
[Router]aaa
[Router-aaa]local-user ftp password cipher hbeutc
Info: Add a new user.
[Router-aaa]local-user ftp ftp-directory flash:
[Router-aaa]local-user ftp service-type ftp
[Router-aaa]local-user ftp privilege level 15
```

配置完成后，在本地创建测试文件 test.test.txt，并设置客户端信息，如图 10.6 所示。设置服务器地址为 211.85.8.1，用户名为 ftp，密码为 hbeutc，然后单击"登录"按钮。

图 10.6　登录服务器

登录成功后，在"本地文件列表"中选择 test.test.txt，单击向右箭头上传文件到 FTP 服务器，结果显示上传成功，如图 10.7 所示。

图 10.7　上传文件

也可通过 dir 命令在 Router 上查看目录下的文件。

```
<Router>dir    //直接查看flash:下的文件
Directory of flash:/
  Idx  Attr     Size(Byte)  Date        Time(LMT)  FileName
    0  drw-              -  Sep 15 2022 08:46:51   dhcp
    1  -rw-              0  Sep 15 2022 12:15:40   test.test.txt
    2  -rw-        121,802  May 26 2014 09:20:58   portalpage.zip
    3  -rw-              0  Sep 15 2022 11:35:13   test.txt
    4  -rw-          2,263  Sep 15 2022 08:46:45   statemach.efs
    5  -rw-        828,482  May 26 2014 09:20:58   sslvpn.zip
    6  -rw-            563  Sep 15 2022 08:46:41   vrpcfg.zip
1,090,732 KB total (784,460 KB free)
```

结果显示，test.test.txt 成功上传至 FTP 服务器 Router。

4. 查看 FTP 服务器日志文件

通过 FTP 服务器日志文件查看所有 FTP 服务器访问记录，如图 10.8 所示。
结果显示，FTP 服务器记录了所有的操作信息。

第 10 章　应用服务器配置

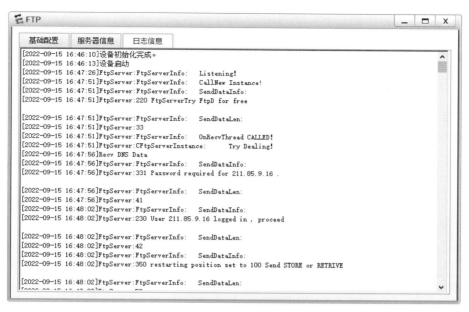

图 10.8　FTP 服务器日志信息

10.2　Web 服务器配置

一、原理概述

Web 服务器是指驻留于互联网上某种类型计算机的程序。当 Web 浏览器（客户端）连接到服务器并请求访问文件时，服务器将处理该请求并将文件反馈到该浏览器上，附带的信息会告诉浏览器如何查看该文件（即文件类型）。

二、实验目的

1. 掌握配置 Web 服务器的方法
2. 掌握 FTP 服务器配合 Web 服务器工作的方法

三、实验内容

本实验模拟企业网络环境，客户机 Client 为客户端设备，需要访问服务器设备 Server。在 Server 上分别创建 Web 服务器和 FTP 服务器，通过 FTP 服务器将 Web 网站文件上传至服务器，服务器自动发布 Web 网站。

四、实验拓扑

Web 服务器配置拓扑如图 10.9 所示。

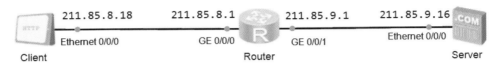

图 10.9 Web 服务器配置拓扑

五、实验地址分配

本实验地址分配如表 10.2 所示。

表 10.2 实验 IP 地址分配表

设备	接口	IP 地址	子网掩码	默认网关
Router（AR2220）	GE 0/0/0	211.85.8.1	255.255.255.0	N/A
	GE 0/0/1	211.85.9.1	255.255.255.0	N/A
Client	Ethernet 0/0/0	211.85.8.18	255.255.255.0	211.85.8.1
Server	Ethernet 0/0/0	211.85.9.16	255.255.255.0	211.85.9.1

六、实验步骤

1. 基本配置

1) Client 的配置

Client 的基本配置如图 10.10 所示。

图 10.10 Client 的基本配置

2) 路由器 Router 的基本配置

```
<Huawei>system-view
[Huawei]sysname Router
[Router]interface GigabitEthernet 0/0/0
[Router-GigabitEthernet0/0/0]ip address 211.85.8.1 24
```

```
[Router-GigabitEthernet0/0/0]interface GigabitEthernet 0/0/1
[Router-GigabitEthernet0/0/1]ip address 211.85.9.1 24
```

3) Server 的基本配置

Server 的基本配置如图 10.11 所示。

图 10.11　Server 的基本配置

4) 测试 Client 与 Server 的连通性。

在 Client 上使用 ping 测试功能测试与 Server 的连通性，在目的 IP 地址中填写 211.85.9.16，次数填写 4，单击"发送"按钮，如图 10.12 所示。

图 10.12　Client 与 Server 连通性测试

结果显示，Client 客户机与 Server 可以通信。

2．创建网页

在本地计算机上创建一个简单的 Web 页面。

在 Notepad（记事本）中，键入下列文本：

```
<html>
<head><title>网站在构建中……</title></head>
<body><h1> 本页面在构建中……</h1>
<p>更多的信息将在这里发布 </p></body>
</html>
```

另存为网页文件 index.html，保存在指定的文件夹中，如 C:\temp。

注意：网站首页一般以 default.html 或 index.html 保存。

3. 配置 FTP 服务器

在 Server 上的"服务器信息"标签中，选择 FtpServer 单选按钮，在"配置"标签中的文件根目录中选择 C:\web 文件夹，单击"启动"按钮启动 FTP 服务器，如图 10.13 所示。

图 10.13 配置 FTP 服务器

在 Client 的"客户端信息"标签中选择 "FtpClient"单选按钮，在服务器地址中填写服务器的 IP 地址 211.85.9.16，用户名为 ftp，密码为 hbeutc。单击"登录"按钮，成功登录 FTP 服务器，如图 10.14 所示。

图 10.14 登录 FTP 服务器

在本地文件列表中选择 C:\temp\index.html，单击向右按钮上传 index.html 到 FTP 服务器上，如图 10.15 所示。

4. 配置 Web 服务器

在 Server 中选择"服务器信息"标签，选中"HttpServer"单选按钮，在文件根目录中选择 C:\Web，刚刚通过 FTP 上传的 index.html 会出现在文本框中。单击"启动"按钮启动 Web 服务器，如图 10.16 所示。

第 10 章 应用服务器配置

图 10.15 上传网页文件

图 10.16 配置 Web 服务器

5．测试

在 Client 上选择"客户端信息"标签，选中"HttpClient"单选按钮，在地址栏里输入 http://211.85.9.16/index.html，单击"获取"按钮，如图 10.17 所示。

```
HTTP/1.1 200 OK        //成功访问
Server: ENSP HttpServer
Auth: HUAWEI
Cache-Control: private
Content-Type: text/html
Content-Length: 165
```

弹出保存 index.html 文件对话框，结果显示，成功访问 Web 页面。

6．查看 Web 服务器日志文件

在 Web 服务器 Server 上，通过 Web 服务器日志信息文件查看所有 Web 服务器访问记录，如图 10.18 所示。

图 10.17　访问 Web 网页

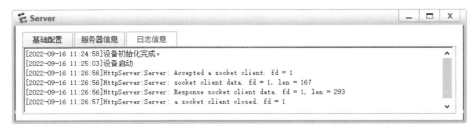

图 10.18　Web　Server 日志信息

10.3　DNS 服务器配置

一、原理概述

域名系统（domain name system，DNS）是进行域名和与之相对应的 IP 地址转换的服务器。DNS 中保存了一张域名和与之相对应的 IP 地址的表，以解析域名。域名是由一串用点分隔的名字组成的，通常包含组织名，而且始终包括两到三个字母的后缀，以指明组织的类型或该域所在的国家或地区。

二、实验目的

1. 掌握配置 DNS 服务器的方法
2. 掌握 DNS 服务器配合 Web 服务器工作的方法

三、实验内容

本实验模拟企业网络环境,企业内部搭建本地 DNS 服务器,为企业各主机提供域名解析服务,设置本地 DNS 服务器,主机 Client1 和 Client2 等可以通过域名 www.abc.com 访问 IP 地址为 211.85.8.18 的 Web 服务器。

四、实验拓扑

DNS 服务器配置拓扑如图 10.19 所示。

图 10.19　DNS 服务器配置拓扑

五、实验地址分配

实验地址分配如表 10.3 所示。

表 10.3　DNS 服务器配置地址分配表

设备	接口	IP 地址	子网掩码	默认网关
Router（AR2220）	GE 0/0/0	211.85.9.254	255.255.255.0	N/A
	GE 0/0/1	211.85.8.254	255.255.255.0	N/A
Client1	Ethernet 0/0/0	211.85.9.16	255.255.255.0	211.85.9.254
Client2	Ethernet 0/0/0	211.85.9.18	255.255.255.0	211.85.9.254
DNS Server	Ethernet 0/0/0	211.85.9.27	255.255.255.0	211.85.9.254
Web Server	Ethernet 0/0/0	211.85.8.18	255.255.255.0	211.85.8.254

六、实验步骤

1. Client 的基本配置

Client1 的基本配置如图 10.20 所示。

图 10.20　Client1 的配置

Client2 的基本配置如图 10.21 所示。

图 10.21　Client2 的配置

2. Router 的基本配置

根据实验地址分配，配置路由器 Router 的基本信息。

```
<Huawei>system-view
[Huawei]sysname Router
[Router]interface GigabitEthernet 0/0/0
[Router-GigabitEthernet0/0/0]ip address 211.85.9.254 24
[Router-GigabitEthernet0/0/0]interface GigabitEthernet 0/0/1
[Router-GigabitEthernet0/0/1]ip address 211.85.8.254 24
```

3. DNS Server 的基本配置

DNS Server 的基本配置如图 10.22 所示。

图 10-22　DNS Server 的基本配置

4. Web Server 的基本配置

Web Server 的基本配置如图 10.23 所示。

第 10 章 应用服务器配置

图 10.23 Web Server 基本配置

5. 测试

基本配置完成后，测试网络连通性。在 Client1 上使用 ping 测试功能测试与 DNS Server 的连通性，如图 10.24 所示。

图 10.24 测试 Client1 与 DNS Server 的连通性

结果显示，Client1 与 DNS Server 可以正常通信。

在 Client1 上使用 **ping 测试**功能测试与 Web Server 的连通性，如图 10.25 所示。

图 10.25 测试 Client1 与 Web Server 的连通性

结果显示，Client1 与 Web Server 可以正常通信。

6. DNS 服务器的配置

Web 服务器的 IP 地址 211.85.8.18 必须与其域名 www.abc.com 绑定。打开 DNS Server 的配置界面，选择"服务器信息"标签，在左侧列表中选择"DNSServer"单选按钮，在右侧配置栏里主机域名输入"www.abc.com"，IP 地址输入"211.85.8.18"，单击"添加"按钮，如图 10.26 所示。

7. Web 服务器的配置

在磁盘 D：上创建 web 文件夹，在 web 文件夹中创建一个网页文件 index.html。打开 Web Server 配置界面，选中"服务器信息"标签，在左侧选择"HttpServer"单选按钮，在右侧配置栏中文件目录选择"D:\web"，单击"启动"按钮，启动 Web 服务器，如图 10.27 所示。

图 10.26　DNS 服务器的配置

图 10.27　Web 服务器的配置

在 Client1 上访问 Web 服务器发布的网页，在 Client1 的配置界面中选择"客户端信息"标签，选择"HttpClient"单选按钮，右侧地址栏里输入"http://www.abc.com"，单击"获取"按钮打开对应网页，访问网页的同时启动，如图 10.28 所示。

图 10.28　客户端使用域名访问 Web 网页

Client1 使用域名访问 Web 服务器同时,交换机 SW1 连接 DNS Server 的接口捕获报文,Client1 与本地域名服务器之间完成一次域名解析请求和响应消息的交互过程,如图 10.29 所示。

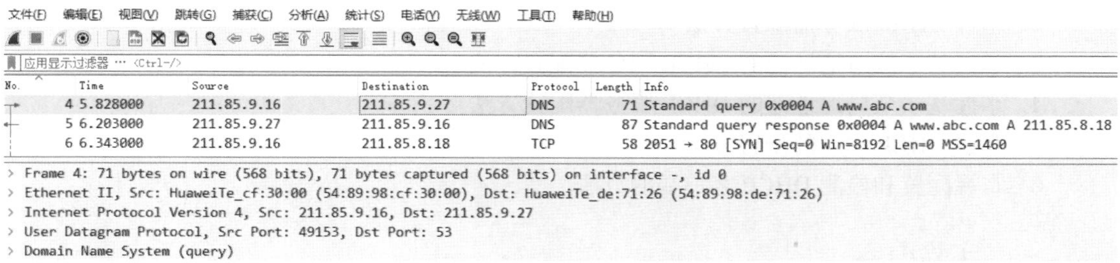

图 10.29　交换机 SW1 连接 DNS　Server 接口捕获报文

结果显示,Client1 向 DNS 服务器请求解析域名 www.abc.com,DNS 服务器解析为 211.85.8.18,之后在 Client1 与 Web 服务器之间建立 TCP 连接。

10.4　DHCP 服务器配置

一、原理概述

动态主机配置协议(dynamic host configuration protocol,DHCP)通常被应用在大型的局域网络环境中,主要作用是集中管理、分配 IP 地址,使网络环境中的主机动态地获得 IP 地址、Gateway 地址、DNS 服务器地址等信息,并能够提升地址的使用率。

DHCP 协议采用 C/S 模型,主机地址的动态分配任务由网络主机驱动。当 DHCP 服务器接收到来自网络主机申请地址的信息时,才会向网络主机发送相关的地址配置等信息,以实现网络主机地址信息的动态配置。

DHCP 具有以下功能:

(1) 保证任何 IP 地址在同一时刻只能由一台 DHCP 客户机所使用;

(2) DHCP 应当可以给用户分配永久固定的 IP 地址;

(3) DHCP 应当可以用其他方法获得 IP 地址的主机共存(如手工配置 IP 地址的主机);

(4) DHCP 服务器应当向现有的 BOOTP 客户端提供服务。

DHCP 服务器的配置有三种方法:

(1) 基于端口地址池的 DHCP 服务器配置;

(2) 基于全局地址池的 DHCP 服务器配置;

(3) 基于中继的 DHCP 服务器配置。

二、实验目的

1. 理解 DHCP 服务器应用场景
2. 掌握 DHCP 服务器配置方法
3. 掌握基于接口地址池的 DHCP 服务器配置方法
4. 掌握基于全局地址池的 DHCP 服务器配置方法
5. 掌握 DHCP 中继的配置方法
6. 掌握配置和检测 DHCP 客户端的方法

三、实验内容

本实验模拟企业网络环境,三个分公司 A、B、C 分别由 Router2、Router3 和 Router1 通过公网路由器 Router4 访问总公司路由器 Router5。公司 A 的路由器 Router2 模拟 DHCP 服务器,采用端口地址池的方式分配 IP 地址。公司 B 的路由器 Router3 模拟 DHCP 服务器,采用全局地址池的方式分配 IP 地址。公司 C 由交换机 SW1 和网关路由器 Router1 通过公网路由器 Router4 访问总公司的 DHCP 服务器 Router5,由于公司 C 与总公司不在同一个网络,需要在 Router1 上配置 DHCP 中继,使公司 C 的主机能跨网络从总部 DHCP 服务器自动获取 IP 地址。

四、实验拓扑

配置 DHCP 服务器的拓扑如图 10.30 所示。

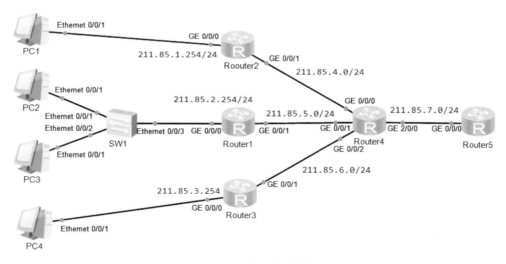

图 10.30 DHCP 服务器配置拓扑

五、实验地址分配

实验地址分配如表 10.4 所示。

第 10 章 应用服务器配置

表 10.4 实验地址分配

设备	接口	IP 地址	子网掩码	默认网关
Router1（AR2220）	GE 0/0/0	211.85.2.254	255.255.255.0	N/A
	GE 0/0/1	211.85.5.1	255.255.255.0	N/A
Router2（AR2220）	GE 0/0/0	211.85.1.254	255.255.255.0	N/A
	GE 0/0/1	211.85.4.1	255.255.255.0	N/A
Router3（AR2220）	GE 0/0/0	211.85.3.254	255.255.255.0	N/A
	GE 0/0/1	211.85.6.1	255.255.255.0	N/A
Router4（AR2220）	GE 0/0/0	211.85.4.2	255.255.255.0	N/A
	GE 0/0/1	211.85.5.2	255.255.255.0	N/A
	GE 0/0/2	211.85.6.2	255.255.255.0	N/A
	GE 2/0/0	211.85.7.1	255.255.255.0	N/A
Router5（AR2220）	GE 0/0/0	211.85.7.2	255.255.255.0	N/A

六、实验步骤

1. 基本配置

根据实验地址分配对各设备进行基本配置，各 PC 自动获取 IP 地址信息，无需配置。

1) 路由器 Router1 的基本配置

```
<Huawei>system-view
[Huawei]sysname Router1
[Router1]interface GigabitEthernet 0/0/0
[Router1-GigabitEthernet0/0/0]ip address 211.85.2.254 24
[Router1-GigabitEthernet0/0/0]interface GigabitEthernet 0/0/1
[Router1-GigabitEthernet0/0/1]ip address 211.85.5.1 24
```

2) 路由器 Router2 的基本配置

```
<Huawei>system-view
[Huawei]sysname Router2
[Router2]interface GigabitEthernet 0/0/0
[Router2-GigabitEthernet0/0/0]ip address 211.85.1.254 24
[Router2-GigabitEthernet0/0/0]interface GigabitEthernet 0/0/1
[Router2-GigabitEthernet0/0/1]ip address 211.85.4.1 24
```

3) 路由器 Router3 的基本配置

```
<Huawei>system-view
[Huawei]sysname Router3
[Router3]interface GigabitEthernet 0/0/0
[Router3-GigabitEthernet0/0/0]ip address 211.85.3.254 24
[Router3-GigabitEthernet0/0/0]interface GigabitEthernet 0/0/1
[Router3-GigabitEthernet0/0/1]ip address 211.85.6.1 24
```

4) 路由器 Router4 的基本配置

因 Router4 接口不够用，先关闭电源，需在 Router4 上添加 1 个 GE 接口卡，如图 10.31 所示。

```
<Huawei>system-view
[Huawei]sysname Router4
[Router4]interface GigabitEthernet 2/0/0
[Router4-GigabitEthernet1/0/0]ip address 211.85.7.1 24
[Router4-GigabitEthernet1/0/0]interface GigabitEthernet 0/0/0
[Router4-GigabitEthernet0/0/0]ip address 211.85.4.2 24
[Router4-GigabitEthernet0/0/0]interface GigabitEthernet 0/0/1
[Router4-GigabitEthernet0/0/1]ip address 211.85.5.2 24
[Router4-GigabitEthernet0/0/1]interface GigabitEthernet 0/0/2
[Router4-GigabitEthernet0/0/2]ip address 211.85.6.2 24
```

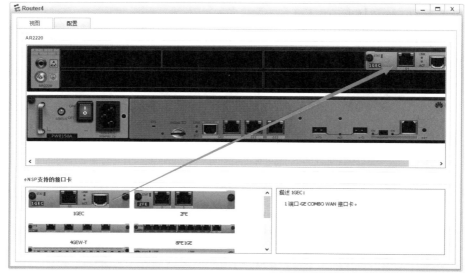

图 10.31　添加 1 个 GE 接口卡

5) 路由器 Router5 的基本配置

```
<Huawei>system-view
[Huawei]sysname Router5
[Router5]interface GigabitEthernet 0/0/0
[Router5-GigabitEthernet0/0/0]ip address 211.85.7.2 24
```

2. 配置 OSPF 网络

在路由器 Router1 上配置 OSPF。

```
[Router1]ospf 1
[Router1-ospf-1]area 0
[Router1-ospf-1-area-0.0.0.0]network 211.85.2.0 0.0.0.255
[Router1-ospf-1-area-0.0.0.0]network 211.85.5.0 0.0.0.255
```

在路由器 Router2 上配置 OSPF。

```
[Router2]ospf 1
[Router2-ospf-1]area 0
[Router2-ospf-1-area-0.0.0.0]network 211.85.1.0 0.0.0.255
[Router2-ospf-1-area-0.0.0.0]network 211.85.4.0 0.0.0.255
```

在路由器 Router3 上配置 OSPF。

```
[Router3]ospf 1
[Router3-ospf-1]area 0
[Router3-ospf-1-area-0.0.0.0]network 211.85.3.0 0.0.0.255
[Router3-ospf-1-area-0.0.0.0]network 211.85.6.0 0.0.0.255
```

在路由器 Router4 上配置 OSPF。

```
[Router4]ospf 1
[Router4-ospf-1]area 0
[Router4-ospf-1-area-0.0.0.0]network 211.85.4.0 0.0.0.255
[Router4-ospf-1-area-0.0.0.0]network 211.85.5.0 0.0.0.255
[Router4-ospf-1-area-0.0.0.0]network 211.85.6.0 0.0.0.255
[Router4-ospf-1-area-0.0.0.0]network 211.85.7.0 0.0.0.255
```

在路由器 Router5 上配置 OSPF。

```
[Router5]ospf 1
[Router5-ospf-1]area 0
[Router5-ospf-1-area-0.0.0.0]network 211.85.7.0 0.0.0.255
```

OSPF 配置完成后，在路由器 Router2 上使用 ping 命令测试与总公司的 Router5 之间的连通性，以测试配置是否正确。其他路由器之间的测试使用同样的方法。

```
<Router2>ping 211.85.7.2
  ping 211.85.7.2: 56  data bytes, press CTRL_C to break
    Reply from 211.85.7.2: bytes=56 Sequence=1 ttl=254 time=30 ms
    Reply from 211.85.7.2: bytes=56 Sequence=2 ttl=254 time=30 ms
    Reply from 211.85.7.2: bytes=56 Sequence=3 ttl=254 time=30 ms
    Reply from 211.85.7.2: bytes=56 Sequence=4 ttl=254 time=40 ms
    Reply from 211.85.7.2: bytes=56 Sequence=5 ttl=254 time=20 ms
  --- 211.85.7.2 ping statistics ---
    5 packet(s) transmitted
    5 packet(s) received
    0.00% packet loss
    round-trip min/avg/max = 20/30/40 ms
```

结果显示，网络能正常通信，OSPF 配置正确。

3. 基于接口地址池的 DHCP 服务器配置

在 Router2 上使用 dhcp enable 命令开启 DHCP 功能，在 Router2 的 GE 0/0/0 接口上配置 dhcp select interface 命令，开启接口的 DHCP 功能，使用 dhcp server lease 命令配置 DHCP 服务器接口地址池中 IP 地址的租期为 2 天（默认租期为 1 天），使用 dhcp server excluded-ip-address 命令指定接口地址池中 211.85.1.1~211.85.1.20 不参与分配，使用 dhcp server dns-list 命令配置 DNS 服务器地址 211.85.1.18，使客户端可以自动获取 DNS 服务器地址。

```
[Router2]dhcp enable
[Router2]interface GigabitEthernet 0/0/0
[Router2-GigabitEthernet0/0/0]dhcp select interface
[Router2-GigabitEthernet0/0/0]dhcp server lease day 2
[Router2-GigabitEthernet0/0/0]dhcp server excluded-ip-address 211.85.1.1 211.85.1.20
[Router2-GigabitEthernet0/0/0]dhcp server dns-list 211.85.1.18
```

在 Router2 上使用 display ip pool 命令查看 DHCP 地址池中的地址分配情况。

第 10 章　应用服务器配置

```
[Router2]display ip pool
-----------------------------------------------------------
  Pool-name          : GigabitEthernet0/0/0
  Pool-No            : 0
  Position           : Interface       Status        : Unlocked
  Gateway-0          : 211.85.1.254
  Mask               : 255.255.255.0
  VPN instance       : --
  IP address Statistic
    Total            :253
    Used             :0           Idle            :233
    Expired          :0           Conflict        :0       Disable    :20
```

结果显示，目前为基于接口的地址池，在 DHCP 地址池中，网关为 211.85.1.254，掩码为 255.255.255.0，IP 地址总数为 253 个，有 20 个地址不参与分配。

在 PC1~PC4 的基础配置界面，在 IPv4 配置中单击"DHCP"单选按钮，如图 10.32 所示。

图 10.32　PC1~PC4 的基本配置

在 PC1 的命令行界面中输入 **ipconfig** 命令查看接口的 IP 地址信息。

```
PC>ipconfig
Link local IPv6 address..........: fe80::5689:98ff:fe97:669a
IPv4 address....................: 211.85.1.253
Subnet mask.....................: 255.255.255.0
Gateway........................: 211.85.1.254
Physical address................: 54-89-98-97-66-9A
DNS server.....................: 211.85.1.18
```

结果显示，PC1 成功获取 IP 地址 211.85.1.253，网关地址 211.85.1.254，DNS 地址 211.85.1.18。
在 PC1 上使用 **ping** 命令测试与总公司路由器 Router5 之间的连通性。

```
PC>ping 211.85.7.2
ping 211.85.7.2: 32 data bytes, Press Ctrl_C to break
From 211.85.7.2: bytes=32 seq=1 ttl=253 time=31 ms
From 211.85.7.2: bytes=32 seq=2 ttl=253 time=31 ms
```

计算机网络实践教程

```
From 211.85.7.2: bytes=32 seq=3 ttl=253 time=16 ms
From 211.85.7.2: bytes=32 seq=4 ttl=253 time=31 ms
From 211.85.7.2: bytes=32 seq=5 ttl=253 time=32 ms
--- 211.85.7.2 ping statistics ---
 5 packet(s) transmitted
 5 packet(s) received
 0.00% packet loss
 round-trip min/avg/max = 16/28/32 ms
```

结果显示，可以通信。表示 PC1 获取的 IP 地址信息使 PC1 可以访问网络。

4. 基于全局地址池的 DHCP 服务器配置

在 Router3 上使用 dhcp enable 命令开启 DHCP 功能；使用 ip pool 命令创建一个全局地址池，名称设置为 hbeutc；使用 network 命令配置全局地址池 hbeutc 可动态分配的网络为 211.85.3.0；使用 lease day 命令配置 DHCP 全局地址池下的地址租期，默认为 1 天；使用 gateway-list 命令为客户端配置网关 211.85.3.254；使用 excluded-ip-address 命令设置 211.85.3.1~211.85.3.20 这些地址不参与自动分配；使用 dns-list 命令为客户端配置 DNS 服务器地址为 211.85.3.18。

```
[Router3]dhcp enable
[Router3]ip pool hbeutc
[Router3-ip-pool-hbeutc]network 211.85.3.0 mask 255.255.255.0
[Router3-ip-pool-hbeutc]lease day 2
[Router3-ip-pool-hbeutc]gateway-list 211.85.3.254
[Router3-ip-pool-hbeutc]excluded-ip-address 211.85.3.1 211.85.3.20
[Router3-ip-pool-hbeutc]dns-list 211.85.3.18
```

在 Router3 的 GE 0/0/0 接口上开启 DHCP 功能，使用 dhcp select global 命令配置该接口采用全局地址池为客户端分配 IP 地址。

```
[Router3]interface GigabitEthernet 0/0/0
[Router3-GigabitEthernet0/0/0]dhcp select global
```

在 Router3 上使用 display ip pool 命令查看地址池信息。

```
[Router3]display ip pool
---------------------------------------------------------------
 Pool-name         : hbeutc
 Pool-No           : 0
 Position          : Local            Status           : Unlocked
 Gateway-0         : 211.85.3.254
 Mask              : 255.255.255.0
 VPN instance      : --
```

```
    IP address Statistic
     Total        :253
     Used         :0           Idle         :233
     Expired      :0           Conflict     :0           Disable    :20
```

结果显示，名为 hbeutc 的地址池有 253 个 IP 地址，使用了 0 个，空闲 233 个，有 20 个地址不参与分配。

在 PC4 的命令行界面中，使用 **ipconfig** 命令查看接口的 IP 地址等信息。

```
PC>ipconfig
Link local IPv6 address...........: fe80::5689:98ff:febc:17b
IPv6 address......................: :: / 128
IPv6 gateway......................: ::
IPv4 address......................: 211.85.3.253
Subnet mask.......................: 255.255.255.0
Gateway...........................: 211.85.3.254
Physical address..................: 54-89-98-BC-01-7B
DNS server........................: 211.85.3.18
```

结果显示，PC4 成功获取 IP 地址 211.85.3.253，网关地址 211.85.3.254，DNS 地址 211.85.3.18。

在 PC4 上使用 **ping** 命令测试 PC4 与 Router5 之间的连通性。

```
PC>ping 211.85.7.2
ping 211.85.7.2: 32 data bytes, Press Ctrl_C to break
From 211.85.7.2: bytes=32 seq=1 ttl=253 time=31 ms
From 211.85.7.2: bytes=32 seq=2 ttl=253 time=31 ms
From 211.85.7.2: bytes=32 seq=3 ttl=253 time=16 ms
From 211.85.7.2: bytes=32 seq=4 ttl=253 time=31 ms
From 211.85.7.2: bytes=32 seq=5 ttl=253 time=16 ms
--- 211.85.7.2 ping statistics ---
 5 packet(s) transmitted
 5 packet(s) received
 0.00% packet loss
 round-trip min/avg/max = 16/25/31 ms
```

结果显示，可以通信。表示 PC4 获取的 IP 地址信息使 PC4 可以访问网络。

5. DHCP 中继的配置

总公司路由器 Router5 配置为 DHCP 服务器，负责为分公司 C 网络分配 IP 地址。在 Router5 上使用 **dhcp enable** 命令启用 DHCP 功能；使用 **ip pool hbeutc** 命令创建名为 hbeutc 的全局地址池，可分配的地址范围为 211.85.2.1~211.85.2.253，默认网关为 211.85.2.254；在 Router5 的 GE 0/0/0 接口上启用 DHCP 功能，指定从全局地址池分配地址。

```
[Router5]dhcp enable
[Router5]ip pool hbeutc
[Router5-ip-pool-hbeutc]network 211.85.2.0 mask 255.255.255.0
[Router5-ip-pool-hbeutc]gateway-list 211.85.2.254
[Router5-ip-pool-hbeutc]interface gigabitEthernet 0/0/0
[Router5-GigabitEthernet0/0/0]dhcp select global
```

配置 Router1 为中继 DHCP，指定 DHCP 服务器为 Router5。配置指定 DHCP 服务器有两种方式：

(1) 直接在 Router1 的 GE 0/0/0 接口下开启 DHCP 中继功能，并直接指定 DHCP 服务器 IP 地址为 211.85.7.2。

```
[Router1]dhcp enable
[Router1]interface GigabitEthernet 0/0/0
[Router1-GigabitEthernet0/0/0]dhcp select relay
[Router1-GigabitEthernet0/0/0]dhcp relay server-ip 211.85.7.2
```

(2) 直接在 Router1 上创建 DHCP 服务器组，指定组名为 dhcp-group，并使用 **dhcp-server** 命令添加远端的 DHCP 服务器 IP 地址 211.85.7.2。在 GE 0/0/0 接口下开启 DHCP 中继功能，并配置所对应的 DHCP 服务器组。

```
[Router5]dhcp server dhcp-group
[Router5]ip pool dhcp-pool
[Router5-dhcp-server-dhcp-group]dhcp-server 211.85.7.2
[Router5-dhcp-server-dhcp-group]interface gigabitEthernet 0/0/0
[Router5-GigabitEthernet0/0/0]dhcp select global
[Router5-GigabitEthernet0/0/0]dhcp relay server-select dhcp-group
```

在 PC2 上使用 **ipconfig** 命令查看地址信息。

```
PC>ipconfig
Link local IPv6 address.............: fe80::5689:98ff:fe29:7868
IPv6 address.......................: :: / 128
IPv6 gateway.......................: ::
IPv4 address.......................: 211.85.2.253
Subnet mask........................: 255.255.255.0
Gateway............................: 211.85.2.254
Physical address...................: 54-89-98-29-78-68
DNS server.........................:
```

结果显示 PC2 从总部 DHCP 服务器获取了 IP 地址 211.85.2.253。
在 PC3 上使用 ipconfig 命令查看地址信息。

```
PC>ipconfig
Link local IPv6 address...........: fe80::5689:98ff:fe6f:5b0b
IPv6 address.....................: :: / 128
IPv6 gateway.....................: ::
IPv4 address.....................: 211.85.2.252
Subnet mask......................: 255.255.255.0
Gateway..........................: 211.85.2.254
Physical address.................: 54-89-98-6F-5B-0B
DNS server.......................:
```

结果显示 PC3 从总部 DHCP 服务器获取了 IP 地址 211.85.2.252。
测试 PC2 与 PC3 之间的连通性。

```
PC>ping 211.85.2.252
ping 211.85.2.252: 32 data bytes, Press Ctrl_C to break
From 211.85.2.252: bytes=32 seq=1 ttl=128 time=47 ms
From 211.85.2.252: bytes=32 seq=2 ttl=128 time=31 ms
From 211.85.2.252: bytes=32 seq=3 ttl=128 time=47 ms
From 211.85.2.252: bytes=32 seq=4 ttl=128 time=31 ms
From 211.85.2.252: bytes=32 seq=5 ttl=128 time=32 ms
--- 211.85.2.252 ping statistics ---
 5 packet(s) transmitted
 5 packet(s) received
 0.00% packet loss
 round-trip min/avg/max = 31/37/47 ms
```

结果显示，PC2 与 PC3 通过 DHCP 中继成功从 DHCP 服务器获得 IP 地址，并能相互通信。
在 PC2 上使用 ping 命令测试与 DHCP 服务器之间的连通性。

```
PC>ping 211.85.7.2
ping 211.85.7.2: 32 data bytes, Press Ctrl_C to break
From 211.85.7.2: bytes=32 seq=1 ttl=253 time=62 ms
From 211.85.7.2: bytes=32 seq=2 ttl=253 time=32 ms
From 211.85.7.2: bytes=32 seq=3 ttl=253 time=31 ms
From 211.85.7.2: bytes=32 seq=4 ttl=253 time=31 ms
From 211.85.7.2: bytes=32 seq=5 ttl=253 time=31 ms
--- 211.85.7.2 ping statistics ---
 5 packet(s) transmitted
 5 packet(s) received
```

```
0.00% packet loss
round-trip min/avg/max = 31/37/62 ms
```

结果显示，通信正常。

整个配置过程，仅在 Router1 上开启 DHCP 中继功能，在分公司 C 网络没有其他 DHCP 配置。由此可见，在网络设计和管理中灵活使用 DHCP 中继功能能够使网络运行更加高效和方便。

10.5　Telnet 服务配置

一、原理概述

Telnet 协议是 TCP/IP 协议族中的一员，是 Internet 远程登录服务的标准协议和主要方式。它为用户提供了在本地计算机上完成控制远程主机的能力。终端使用 telnet 程序连接到服务器。终端可以在 telnet 程序中输入命令，这些命令会在服务器上运行，就像直接在服务器的控制台上输入一样，在本地就能控制服务器。要开始一个 telnet 会话，必须输入用户名和密码来登录服务器。Telnet 是常用的远程控制 Web 服务器的方法。用户可以通过 Telnet 的方式在一台设备上对多台（本地或远程）设备进行管理或配置。

Telnet 远程登录服务分为以下 4 个过程：

(1) 本地与远程主机建立连接。该过程实际上是建立一个 TCP 连接，用户必须知道远程主机的 IP 地址或域名。

(2) 将本地终端上输入的用户名和口令及以后输入的任何命令或字符以 NVT（net virtual terminal）格式传送到远程主机。该过程实际上是从本地主机向远程主机发送一个 IP 数据包。

(3) 将远程主机输出的 NVT 格式的数据转化为本地所能接受的格式送回本地终端，包括输入命令回显和命令执行结果。

(4) 最后，本地终端对远程主机进行撤销连接。该过程是撤销一个 TCP 连接。

二、实验目的

1. 理解 Telnet 的应用场景
2. 掌握 Telnet 的基本配置方法
3. 掌握 Telnet 密码验证的配置
4. 掌握 Telnet 用户级别的修改方法

三、实验内容

本实验模拟企业网络环境，路由器 Router1 是总公司网络中心的核心设备。公司员工在家或是在外地出差，Router2 模拟在家办公员工的主机，Router3 模拟外地出差员工的主机，Router4 模拟 Internet 连接设备。现需要在 Router1 上配置 Telnet 服务，使在家办公员工和外地出差员工的主机能远程管理机房设备。为了提高安全性，Telnet 需要使用密码验证，只有网络管理员能对设备进行配置和管理，普通用户只能监控设备。

四、实验拓扑

Telnet 服务配置拓扑如图 10.33 所示。

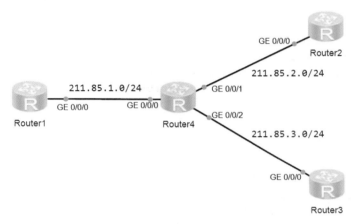

图 10.33　Telnet 服务配置拓扑

五、实验地址分配

本实验地址分配如表 10.5 所示。

表 10.5　实验地址分配

设备	接口	IP 地址	子网掩码	默认网关
Router1（AR2220）	GE 0/0/0	211.85.1.1	255.255.255.0	N/A
Router2（AR2220）	GE 0/0/0	211.85.2.1	255.255.255.0	N/A
Router3（AR2220）	GE 0/0/0	211.85.3.1	255.255.255.0	N/A
Router4（AR2220）	GE 0/0/0	211.85.1.2	255.255.255.0	N/A
	GE 0/0/1	211.85.2.2	255.255.255.0	N/A
	GE 0/0/2	211.85.3.2	255.255.255.0	N/A

六、实验步骤

1．基本配置

根据实验地址分配对各设备进行基本配置，各 PC 自动获取 IP 地址信息，无需配置。

1) 路由器 Router1 的基本配置

```
<Huawei>system-view
[Huawei]sysname Router1
[Router1]interface GigabitEthernet 0/0/0
[Router1-GigabitEthernet0/0/0]ip address 211.85.1.1 24
```

2) 路由器 Router2 的基本配置

```
<Huawei>system-view
[Huawei]sysname Router2
[Router2]interface GigabitEthernet 0/0/0
[Router2-GigabitEthernet0/0/0]ip address 211.85.2.1 24
```

3) 路由器 Router3 的基本配置

```
<Huawei>system-view
[Huawei]sysname Router3
[Router3]interface GigabitEthernet 0/0/0
[Router3-GigabitEthernet0/0/0]ip address 211.85.3.1 24
```

4) 路由器 Router4 的基本配置

```
<Huawei>system-view
[Huawei]sysname Router4
[Router4]interface GigabitEthernet 0/0/0
[Router4-GigabitEthernet0/0/0]ip address 211.85.1.2 24
[Router4-GigabitEthernet0/0/0]interface GigabitEthernet 0/0/1
[Router4-GigabitEthernet0/0/1]ip address 211.85.2.2 24
[Router4-GigabitEthernet0/0/1]interface GigabitEthernet 0/0/2
[Router4-GigabitEthernet0/0/2]ip address 211.85.3.2 24
```

2. 配置 OSPF 网络

在路由器 Router1 上配置 OSPF。

```
[Router1]ospf 1
[Router1-ospf-1]area 0
```

第 10 章 应用服务器配置

```
[Router1-ospf-1-area-0.0.0.0]network 211.85.1.0 0.0.0.255
```

在路由器 Router2 上配置 OSPF。

```
[Router2]ospf 1
[Router2-ospf-1]area 0
[Router2-ospf-1-area-0.0.0.0]network 211.85.2.0 0.0.0.255
```

在路由器 Router3 上配置 OSPF。

```
[Router3]ospf 1
[Router3-ospf-1]area 0
[Router3-ospf-1-area-0.0.0.0]network 211.85.3.0 0.0.0.255
```

在路由器 Router4 上配置 OSPF。

```
[Router4]ospf 1
[Router4-ospf-1]area 0
[Router4-ospf-1-area-0.0.0.0]network 211.85.1.0 0.0.0.255
[Router4-ospf-1-area-0.0.0.0]network 211.85.2.0 0.0.0.255
[Router4-ospf-1-area-0.0.0.0]network 211.85.3.0 0.0.0.255
```

在 Router2 上使用 ping 命令测试与 Router1 之间的连通性。

```
<Router2>ping 211.85.1.1
  ping 211.85.1.1: 56  data bytes, press CTRL_C to break
    Reply from 211.85.1.1: bytes=56 Sequence=1 ttl=254 time=30 ms
    Reply from 211.85.1.1: bytes=56 Sequence=2 ttl=254 time=40 ms
    Reply from 211.85.1.1: bytes=56 Sequence=3 ttl=254 time=30 ms
    Reply from 211.85.1.1: bytes=56 Sequence=4 ttl=254 time=30 ms
    Reply from 211.85.1.1: bytes=56 Sequence=5 ttl=254 time=30 ms
  --- 211.85.1.1 ping statistics ---
    5 packet(s) transmitted
    5 packet(s) received
    0.00% packet loss
    round-trip min/avg/max = 30/32/40 ms
```

结果显示，OSPF 路由表配置成功。可以用同样的方法测试 Router3 与 Router1 之间的连通性。

3. 配置 Telnet 的认证模式

在 Router1 上使用 user-interface vty 0 4 命令进入虚拟终端，使用 authentication-mode password 命令配置认证模式为密码认证（加密的方式显示密码），并设置密码为 hbeutc。

```
[Router1]user-interface vty 0 4
[Router1-ui-vty0-4]authentication-mode password
Please configure the login password (maximum length 16):hbeutc
```

在代表在家办公员工的主机的路由器 Router2 上使用 Telnet 连接 Router1。

```
<Router2>telnet 211.85.1.1
 Press CTRL_] to quit telnet mode
 Trying 211.85.1.1 ...
 Connected to 211.85.1.1 ...
 Login authentication
 Password:              //输入密码不显示，输入正确进入远程设备
<Router1>
```

结果显示，远程登录成功，在 Router2 上进入 Router1 的用户视图。

在代表在外出差员工的主机的路由器 Router3 上使用 telnet 命令连接 Router1。

```
<Router3>telnet 211.85.1.1
 Press CTRL_] to quit telnet mode
 Trying 211.85.1.1 ...
 Connected to 211.85.1.1 ...
Login authentication
Password:
<Router1>
```

结果显示，远程登录成功，在 Router3 上进入 Router1 的用户视图。

Router2 和 Router3 登录成功后，在 Router1 上使用 display users 命令查看已登录的用户信息。

```
[Router1]display users
  User-Intf  Delay    Type  Network Address  AuthenStatus  AuthorcmdFlag
+ 0  CON 0   00:00:00                        pass
  Username : Unspecified
  129 VTY 0  00:00:05  TEL  211.85.2.1       pass
  Username : Unspecified
  130 VTY 1  00:00:45  TEL  211.85.3.1       pass
```

```
Username : Unspecified
```

结果显示，Router2 和 Router3 成功登录 Router1。

4. 配置 Telnet 区分不同用户的权限

根据企业需求，管理员拥有设备的配置和管理权限，普通员工只能拥有设备的监控权限。默认情况下，VTY 用户界面的用户级别为 0（参观级），只能使用 **ping**、**tracert** 等网络测试命令。

在 Router1 上配置 Telnet 的用户级别为 1（监控级）。普通员工使用密码登录设备，只能使用 **display** 等命令监控设备。

```
[Router1]user-interface vty 0 4
[Router1-ui-vty0-4]authentication-mode password
Please configure the login password (maximum length 16):user
[Router1-ui-vty0-4]user privilege level 1
```

在 Router2 上模拟普通用户登录到 Router1。

```
<Router2>telnet 211.85.1.1
  Press CTRL_] to quit telnet mode
  Trying 211.85.1.1 ...
  Connected to 211.85.1.1 ...
Login authentication
Password:                    //输入密码user
<Router1>system-view
        ^
Error: Unrecognized command found at '^' position.
```

结果显示，登录后，无法从用户视图切换到系统视图，这是因为用户级别不够。

管理员使用单独的用户名和密码登录设备，拥有监控级权限。将 VTY 用户界面的认证模式修改成 AAA 认证，这样才能使用本地的用户名和密码进行认证。默认情况下，设备的 AAA 认证功能是开启的，只需要为管理员在本地配置相应的用户名和密码即可。

在 Router1 上使用 **aaa** 命令进入 AAA 视图，在 AAA 视图下配置本地用户名 admin 和密码 admin123，并且将该用户的级别修改为 3（管理级）。

```
[Router1]aaa
[Router1-aaa]local-user admin service-type telnet
[Router1-aaa]local-user admin password cipher admin123 privilege level 3
[Router1-aaa]local-user admin service-type telnet
[Router1-aaa]user-interface vty 0 4
```

```
[Router1-ui-vty0-4]authentication-mode aaa
```

将 Router3 模拟成管理员主机，使用用户名 admin 和密码 admin123 远程登录到 Router1。正确登录后进入 Router1 的用户视图，在用户视图下输入 **system-view** 命令进入 Router1 的系统视图，从而对 Router1 进行所有相关的配置和管理操作。

```
<Router3>telnet 211.85.1.1
  Press CTRL_] to quit telnet mode
  Trying 211.85.1.1 ...
  Connected to 211.85.1.1 ...
Login authentication
Username:admin
Password:            //输入密码admin123
------------------------------------------------------------------
  User last login information:
------------------------------------------------------------------
  Access Type: Telnet
  IP-Address : 211.85.3.1
  Time       : 2022-09-25 02:02:06-08:00
------------------------------------------------------------------
<Router1>system-view
Enter system view, return user view with Ctrl+Z.
[Router1]
```

结果显示，Router3 使用 **telnet** 命令远程登录 Router1，拥有管理员权限。

10.6 SSH 配置

一、原理概述

传统的网络服务 Telnet 和 FTP 等在本质上都是不安全的，因为它们在网络上用明文传送口令和数据，口令和数据容易被截获。这些服务的安全验证方式容易受到"中间人"方式的攻击。

SSH 为 Secure Shell 的缩写，是建立在应用层基础上的安全协议。SSH 是较可靠、专为远程登录会话和其他网络服务提供安全性的协议。利用 SSH 协议可以有效防止远程管理过程中的信息泄露问题。

STelnet 为 SSH Telnet 的缩写，在一个不安全的网络环境中，服务器通过 SSH 对客户端的认证及双向的数据加密，为网络终端访问提供安全的 Telnet 服务。

第 10 章 应用服务器配置

二、实验目的

1. 理解 SSH 的基本原理
2. 掌握配置 SSH Password 认证的方法
3. 掌握 SFTP 的配置

三、实验内容

本实验模拟企业网络环境，路由器 Router1 模拟分公司主机（eNSP 中的 PC 没有 SSH 客户端），作为 SSH 的 Client。现需要在 Router1 上配置 Telnet 服务，路由器 Router2 是总公司网络中心的核心设备，作为 SSH 的 Server。远程主机 Router1 通过 SSH 协议远程登录到路由器 Router2 上进行各种配置。

四、实验拓扑

本实验配置拓扑如图 10.34 所示。

图 10.34　SSH 配置拓扑

五、实验地址分配

本实验地址分配如表 10.6 所示。

表 10.6　SSH 配置地址分配表

设备	接口	IP 地址	子网掩码	默认网关
Router1（AR2220）	GE 0/0/0	211.85.1.1	255.255.255.0	N/A
Router2（AR2220）	GE 0/0/0	211.85.1.2	255.255.255.0	N/A

六、实验步骤

1. 基本配置

根据实验地址分配对各设备进行基本配置。

1）路由器 Router1 的基本配置

```
<Huawei>system-view
[Huawei]sysname Router1
```

```
[Router1]interface GigabitEthernet 0/0/0
[Router1-GigabitEthernet0/0/0]ip address 211.85.1.1 24
```

2) 路由器 Router2 的基本配置

```
<Huawei>system-view
[Huawei]sysname Router2
[Router2]interface GigabitEthernet 0/0/0
[Router2-GigabitEthernet0/0/0]ip address 211.85.1.2 24
```

3) 测试 Router1 与 Route2 之间的连通性

```
<Router1>ping 211.85.1.2
 ping 211.85.1.2: 56  data bytes, press CTRL_C to break
   Reply from 211.85.1.2: bytes=56 Sequence=1 ttl=255 time=100 ms
   Reply from 211.85.1.2: bytes=56 Sequence=2 ttl=255 time=20 ms
   Reply from 211.85.1.2: bytes=56 Sequence=3 ttl=255 time=20 ms
   Reply from 211.85.1.2: bytes=56 Sequence=4 ttl=255 time=20 ms
   Reply from 211.85.1.2: bytes=56 Sequence=5 ttl=255 time=20 ms
 --- 211.85.1.2 ping statistics ---
   5 packet(s) transmitted
   5 packet(s) received
   0.00% packet loss
   round-trip min/avg/max = 20/36/100 ms
```

结果显示，Router1 与 Router2 之间可以通信。

2. 配置 SSH Server

SSH 协议支持对报文压缩并加密传输，因此在通过不安全的网络进行远程登录中，建议使用 SSH 协议。

成功完成 SSH 登录的首要条件是配置并产生本地 RSA 密钥对，在进行其他 SSH 配置之前先要生成本地密钥对，生成的密钥对将保存在设备中，重启后不会丢失。

在 Router2 上使用 **rsa local-key-pair create** 命令生成密钥对，公钥名称一般默认为 Host，如设备已经存在该名称的密钥，可以输入 y 覆盖密钥。

```
[Router2]rsa local-key-pair create
 The key name will be: Host        //自动命名为Host
 % RSA keys defined for Host already exist.
 Confirm to replace them? (y/n)[n]:y        //如果不存在直接创建
 The range of public key size is (512 ~ 2048).
 NOTES: If the key modulus is greater than 512,
```

```
            It will take a few minutes.
 Input the bits in the modulus[default = 512]:   //直接回车，默认为512
 Generating keys.................................................++++++++++++
 .........++++++++++..++++++++...........+++++++
```

配置完成后在 Router2 上用 **display rsa local-key-pair public** 命令查看本地密钥对中的公钥部分信息。

```
[Router2]display rsa local-key-pair public
=====================================================
Time of Key pair created: 2022-09-25 12:59:48-08:00
Key name: Host
Key type: RSA encryption Key
=====================================================
Key code:
3047
  0240
    B92F3110 67914D74 2A6A5915 3AC784B5 2AD6FF80
    ......
  0203
    010001

=====================================================
Time of Key pair created: 2022-09-25 12:59:49-08:00
Key name: Server
Key type: RSA encryption Key
=====================================================
Key code:
3067
  0260
    B6713D12 264073D1 69C46451 0FD7124A CFCAB6E3
    ......
  0203
    010001
```

结果显示，已经生成了本地 RSA 主机密钥对，包括密钥对产生的时间、公钥的名称和公钥的类型。

在 Router2 上配置 VTY 用户界面，设置用户的验证方式为 AAA 授权验证方式；指定 VTY 类型用户界面只支持 SSH 协议，设备将自动禁止 Telnet 功能；使用 **local-user** 命令创建本地用户和用户口令，并以密文方式显示用户口令，指定用户名为 admin，密码为 admin123；使用 **local-user admin service-type ssh** 命令配置本地用户的接入类型为 SSH；使用 **ssh user** 命令创建 SSH 用户，用户名为 admin，指定 SSH 用户的认证方式为 password，即密码认证方式。

```
[Router2]user-interface vty 0 4
[Router2-ui-vty0-4]authentication-mode aaa
[Router2-ui-vty0-4]protocol inbound ssh
[Router2-ui-vty0-4]quit
[Router2]aaa
[Router2-aaa]local-user admin password cipher admin123
[Router2-aaa]local-user admin service-type ssh
[Router2-aaa]quit
[Router2]ssh user admin authentication-type password
```

此处还可以继续使用 local-user admin privilege level 命令配置本地用户的优先级。其取值范围为 0~15，取值越大，代表用户的优先级越高。不同级别的用户登录后，只能使用等于或低于自身级别的命令，默认值为 3，代表管理级。

默认情况下，设备的 SSH 服务器功能为关闭状态，只有开启此功能，客户端才能以 SSH 方式与设备建立连接。在 Router2 上开启设备的 SSH 功能，使用 **display ssh user-information admin** 命令在 SSH 服务器查看 SSH 用户的配置信息。如果不在命令末尾指定 SSH 用户，则可以查看 SSH 服务器上所有的 SSH 用户配置信息。

```
[Router2]stelnet server enable
Info: Succeeded in starting the STELNET server.
[Router2]display ssh user-information admin
--------------------------------------------------------
 Username         Auth-type          User-public-key-name
--------------------------------------------------------
 admin            password           null
--------------------------------------------------------
```

结果显示 SSH 用户的用户名为 admin，认证方式为 password。

在 Router2 上使用 **display ssh server status** 命令，可以查看 SSH 服务器全局配置信息。

```
[Router2]display ssh server status
SSH version                            :1.99
 SSH connection timeout                :60 seconds
 SSH server key generating interval    :0 hours
 SSH Authentication retries            :3 times
 SFTP Server                           :Disable
 Stelnet server                        :Enable
```

结果显示，在 Router2 上 STelnet 状态为启用状态。

3. 配置 SSH Client

当 SSH 客户端第一次登录 SSH 服务器时，客户端没有保存 SSH 服务器的 RSA 公钥，会对服务器的 RSA 有效性公钥检查，登录服务器失败。当客户端 Router1 第一次登录时，需要开启 SSH 客户端首次认证功能，不对 SSH 服务器的 RSA 公钥进行有效性检查。

在 Router1 上使用 **ssh client first-time enable** 命令启用客户端认证功能，使用 **stelnet** 命令连接 SSH 服务器，输入用户名 admin。

```
[Router1]ssh client first-time enable
[Router1]stelnet 211.85.1.2
Please input the username:admin
Trying 211.85.1.2 ...
Press CTRL+K to abort
Connected to 211.85.1.2 ...
The server is not authenticated. Continue to access it? (y/n)[n]:y
Sep 25 2022 15:58:42-08:00 Router1 %%01SSH/4/CONTINUE_KEYEXCHANGE(l)[0]:The
server had not been authenticated in the process of exchanging keys. When deciding
whether to continue, the user chose Y.
[Router1]
Save the server's public key? (y/n)[n]:y
The server's public key will be saved with the name 211.85.1.2. Please wait...
Sep 25 2022 15:58:52-08:00 Router1 %%01SSH/4/SAVE_PUBLICKEY(l)[1]:When
deciding whether to save the server's public key 211.85.1.2, the user chose Y.
[Router1]
Enter password:         //输入密码：admin123
<Router2>               //远程登录到Router2
```

结果显示，第一次登录时，启动了 SSH 客户端首次认证功能，登录后，系统将自动分配并保存 RSA 公钥。在 Router1 上再次登录 SSH 服务器，用户名为 admin，密码为 admin123。

```
[Router1]stelnet 211.85.1.2
Please input the username:admin
Trying 211.85.1.2 ...
Press CTRL+K to abort
Connected to 211.85.1.2 ...
Enter password:
    ------------------------------------------------------------
    User last login information:
    ------------------------------------------------------------
    Access Type: SSH
    IP-Address : 211.85.1.1 ssh
    Time       : 2022-09-25 15:59:08-08:00
    ------------------------------------------------------------
```

结果显示，登录密码输入后，远程登录 Router2 成功。

在 Router2 上使用 **display ssh server session** 命令查看 SSH 服务器的当前会话连接。

```
[Router2]display ssh server session
-----------------------------------------------------------------
 Conn   Ver   Encry    State   Auth-type        Username
-----------------------------------------------------------------
 VTY 0  2.0   AES      run     password         admin
-----------------------------------------------------------------
```

结果显示，客户端 Router1 成功连接到 SSH 服务器 Router2。

4. 配置 SFTP Server

在路由器 Router2 上配置 SFTP Server，使用 **aaa** 命令进入 AAA 视图，创建一个 ftp 用户，配置密码为 ftp123，以密文方式显示；使用 **local-user ftp privilege level 3** 命令配置本地用户的优先级为 3，即管理级；使用 **local-user ftp ftp-directory** 命令为用户指定可访问的文件夹，默认为空，如果不配置，用户将无法登录；使用 **ssh user** 命令新建 SSH 用户，用户名为 ftp，指定用户的认证方式为 password，即密码认证方式；使用 **sftp server enable** 命令开启 SFTP 服务器功能。

```
[Router2]aaa
[Router2-aaa]local-user ftp password cipher ftp123
 Info: Add a new user.
[Router2-aaa]local-user ftp service-type ssh
[Router2-aaa]local-user ftp privilege level 3
[Router2-aaa]local-user ftp ftp-directory flash:
[Router2-aaa]quit
[Router2]ssh user ftp authentication-type password
 Authentication type setted, and will be in effect next time
[Router2]sftp server enable
 Info: Succeeded in starting the SFTP server.
```

结果显示，成功启动 SFTP Server。

配置完成后，使用 **display ssh server status** 命令查看 SSH 服务器的配置信息。

```
[Router2]display ssh server status
 SSH version                             :1.99
 SSH connection timeout                  :60 seconds
 SSH server key generating interval      :0 hours
 SSH Authentication retries              :3 times
 SFTP Server                             :Enable
 Stelnet server                          :Enable
```

第 10 章　应用服务器配置

结果显示，SFTP 服务已经开启。

在客户端 Router1 上使用 **sftp** 命令连接 SSH 服务器，并输入用户名 ftp 和口令 ftp123。

```
[Router1]sftp 211.85.1.2
Please input the username:ftp
Trying 211.85.1.2 ...
Press CTRL+K to abort
Enter password:
sftp-client>
```

结果显示成功登录。

在 Router2 上使用 display ssh server session 命令查看 SSH 会话连接信息。

```
[Router2]display ssh server session
---------------------------------------------------------------
 Conn   Ver   Encry    State  Auth-type       Username
---------------------------------------------------------------
 VTY 0  2.0   AES      run    password        ftp
---------------------------------------------------------------
```

结果显示，客户端 Router1 已经成功连接到 SSH 服务器，可以进行各种配置。

第 11 章
VPN 配置

11.1　IPSec 基本配置

一、原理概述

互联网安全协议（Internet protocol security，IPSec）是一个协议包，通过对 IP 协议的分组进行加密和认证来保护 IP 协议的网络传输协议簇。IPSec 提供端到端通信安全，由终端计算机完成安全操作。这种模式可以用来构建虚拟专用网（VPN），而这也是 IPSec 最主要的用途之一。

IPSec 可以实现以下四项功能。

(1) 数据机密性：IPSec 发送方将包加密后再通过网络发送。

(2) 数据完整性：IPSec 可以验证 IPSec 发送方发送的包，以确保数据传输时没有被改变。

(3) 数据认证：IPSec 接受方能够鉴别 IPsec 包的发送起源。此服务依赖数据的完整性。

(4) 反重放：IPSec 接受方能检查并拒绝重放包。

二、实验目的

1. 理解 IPSec 的基本原理
2. 掌握配置 IPSec 的方法
3. 掌握 SFTP 的配置

三、实验内容

本实验模拟企业网络环境，路由器 Router1 模拟总公司主机（eNSP 中的 PC 没有 SSH 客户端），作为 SSH 的 Client。现需要在 Router1 上配置 Telnet 服务，路由器 Router2 是分公司网络中心的核心设备,作为 SSH 的 Server。远程主机 Router2 通过 SSH 协议远程登录到路由器 Router1 上进行各种配置。

四、实验拓扑

IPSec 基本配置拓扑如图 11.1 所示。

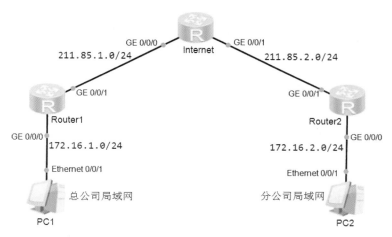

图 11.1　IPSec 基本配置拓扑

五、实验地址分配

实验地址分配如表 11.1 所示。

表 11.1　实验地址分配

设备	接口	IP 地址	子网掩码	默认网关
Router1（AR2220）	GE 0/0/0	172.16.1.1	255.255.255.0	N/A
	GE 0/0/1	211.85.1.1	255.255.255.0	N/A
Router2（AR2220）	GE 0/0/0	172.16.2.1	255.255.255.0	N/A
	GE 0/0/1	211.85.2.2	255.255.255.0	N/A
Internet（AR2220）	GE 0/0/0	211.85.1.2	255.255.255.0	N/A
	GE 0/0/1	211.85.2.1	255.255.255.0	N/A
PC1	Ethernet 0/0/1	172.16.1.2	255.255.255.0	172.16.1.1
PC2	Ethernet 0/0/1	172.16.2.2	255.255.255.0	172.16.2.1

六、实验步骤

1. 基本配置

根据实验地址分配对各设备进行基本配置。

1) PC1 的配置

PC1 的基本配置如图 11.2 所示。

图 11.2　PC1 的基本配置

2) PC2 的配置

PC2 的基本配置如图 11.3 所示。

图 11.3　PC2 的基本配置

3) Router1 的基本配置

```
<Huawei>system-view
[Huawei]sysname Router1
[Router1]interface GigabitEthernet 0/0/0
[Router1-GigabitEthernet0/0/0]ip address 172.16.1.1 24
[Router1-GigabitEthernet0/0/0]interface GigabitEthernet 0/0/1
[Router1-GigabitEthernet0/0/1]ip address 211.85.1.1 24
```

4) Router2 的基本配置

```
<Huawei>system-view
[Huawei]sysname Router2
[Router2]interface GigabitEthernet 0/0/0
[Router2-GigabitEthernet0/0/0]ip address 172.16.2.1 24
```

第 11 章 VPN 配置

```
[Router2-GigabitEthernet0/0/0]interface GigabitEthernet 0/0/1
[Router2-GigabitEthernet0/0/1]ip address 211.85.2.2 24
```

5) 路由器 Internet 的基本配置

```
<Huawei>system-view
[Huawei]sysname Internet
[Internet]interface GigabitEthernet 0/0/0
[Internet-GigabitEthernet0/0/0]ip address 211.85.1.2 24
[Internet-GigabitEthernet0/0/0]interface GigabitEthernet 0/0/1
[Internet-GigabitEthernet0/0/1]ip address 211.85.2.1 24
```

2. OSPF 网络配置

1) Router1 的 OSPF 配置

```
[Router1]ospf 1
[Router1-ospf-1]area 0
[Router1-ospf-1-area-0.0.0.0]network 172.16.1.0 0.0.0.255
[Router1-ospf-1-area-0.0.0.0]network 211.85.1.0 0.0.0.255
```

2) Router2 的 OSPF 配置

```
[Router2]ospf 1
[Router2-ospf-1]area 0
[Router2-ospf-1-area-0.0.0.0]network 172.16.2.0 0.0.0.255
[Router2-ospf-1-area-0.0.0.0]network 211.85.2.0 0.0.0.255
```

3) 路由器 Internet 的 OSPF 配置

```
[Internet]ospf 1
[Internet-ospf-1]area 0
[Internet-ospf-1-area-0.0.0.0]network 211.85.1.0 0.0.0.255
[Internet-ospf-1-area-0.0.0.0]network 211.85.2.0 0.0.0.255
```

4) 测试

在 PC1 上使用 ping 命令测试与 PC2 之间的连通性。

计算机网络实践教程

```
PC>ping 172.16.2.2
From 172.16.2.2: bytes=32 seq=2 ttl=125 time=31 ms
From 172.16.2.2: bytes=32 seq=2 ttl=125 time=31 ms
From 172.16.2.2: bytes=32 seq=3 ttl=125 time=16 ms
From 172.16.2.2: bytes=32 seq=4 ttl=125 time=31 ms
From 172.16.2.2: bytes=32 seq=5 ttl=125 time=31 ms
--- 172.16.2.2 ping statistics ---
  5 packet(s) transmitted
  5 packet(s) received
  20.00% packet loss
  round-trip min/avg/max = 0/27/31 ms
```

结果显示，可以通信，说明 OSPF 网络配置正确。

3. 定义各自要保护的数据流

在 Router1 上配置高级 ACL，定义由网络 172.16.1.0/24 到网络 172.16.2.0/24 的数据流。

```
[Router1]acl number 3001
[Router1-acl-adv-3001]rule permit ip source 172.16.1.0 0.0.0.255
destination 172.16.2.0 0.0.0.255
```

在 Router2 上配置高级 ACL，定义由网络 172.16.2.0/24 到网络 172.16.1.0/24 的数据流。

```
[Router2]acl number 3001
[Router2-acl-adv-3001]rule permit ip source 172.16.2.0 0.0.0.255
destination 172.16.1.0 0.0.0.255
```

4. 创建 IPSec 安全提议

在 Router1 上使用 ipsec proposal 命令创建名为 tran1 的安全提议，使用 transform 命令配置采用的安全协议为 ESP，使用 esp authentication-algorithm 命令配置 ESP 的认证算法为 md5，使用 esp encryption-algorithm 3des 命令配置 ESP 的加密算法为 3des。

```
[Router1]ipsec proposal tran1
[Router1-ipsec-proposal-tran1]transform esp
[Router1-ipsec-proposal-tran1]esp authentication-algorithm md5
[Router1-ipsec-proposal-tran1]esp encryption-algorithm 3des
```

在 Router2 上配置 IPSec 安全提议。

第 11 章 VPN 配置

```
[Router2]ipsec proposal tran1
[Router2-ipsec-proposal-tran1]transform esp
[Router2-ipsec-proposal-tran1]esp authentication-algorithm md5
[Router2-ipsec-proposal-tran1]esp encryption-algorithm 3des
```

配置完成后，在 Router1 上使用 display ipsec proposal 显示所配置的信息。

```
<Router1>display ipsec proposal
Number of proposals: 1
IPSec proposal name: tran1
 Encapsulation mode: Tunnel
 Transform          : esp-new
 ESP protocol       : Authentication MD5-HMAC-96
                      Encryption     3DES
```

在 Router2 上使用 display ipsec proposal 显示所配置的信息。

```
<Router2>display ipsec proposal
Number of proposals: 1
IPSec proposal name: tran1
 Encapsulation mode: Tunnel
 Transform          : esp-new
 ESP protocol       : Authentication MD5-HMAC-96
                      Encryption     3DES
```

5. 配置 IKE 对等体

在 Router1 上使用 ike proposal 5 命令配置 IKE 安全提议，使用 dh 命令配置 IKE 协商时使用的 DH 组为 group14。

```
[Router1]ike proposal 5
[Router1-ike-proposal-5]encryption-algorithm 3des
[Router1-ike-proposal-5]authentication-algorithm md5
[Router1-ike-proposal-5]dh group14
```

在 Router1 上使用 ike peer 命令配置 IKE 对等体，使用 ike-proposal 命令绑定 ike proposal，使用 pre-shared-key 命令指定共享秘钥为 hbeutc，使用 remote-address 命令设定对端用于接收 IPSec 报文的 IP 地址。

```
[Router1]ike peer abc v1
[Router1-ike-peer-abc]ike-proposal 5
[Router1-ike-peer-abc]pre-shared-key cipher hbeutc
[Router1-ike-peer-abc]remote-address 211.85.2.2
```

在 Router2 上配置 IKE 安全提议。

```
[Router2]ike proposal 5
[Router2-ike-proposal-5]encryption-algorithm 3des
[Router2-ike-proposal-5]authentication-algorithm md5
[Router2-ike-proposal-5]dh group14
```

在 Router2 上配置 IKE 对等体，并根据默认配置，配置预共享密钥和对端 IP 地址。

```
[Router2]ike peer bca v1
[Router2-ike-peer-bca]ike-proposal 5
[Router2-ike-peer-bca]pre-shared-key cipher hbeutc
[Router2-ike-peer-bca]remote-address 211.85.1.1
```

6. 创建安全策略

在 Router1 上使用 ipsec policy 命令创建一条动态协商方式安全策略，名称为 map1，序列号为 10。在该安全策略视图下分别指定引用 ike 对等体 abc、安全提议 tran1 和 ACL 3001。

```
[Router1]ipsec policy map1 10 isakmp
[Router1-ipsec-policy-isakmp-map1-10]ike-peer abc
[Router1-ipsec-policy-isakmp-map1-10]proposal tran1
[Router1-ipsec-policy-isakmp-map1-10]security acl 3001
```

在 Router2 上配置 IKE 动态协商方式安全策略。

```
[Router2]ipsec policy use1 10 isakmp
[Router2-ipsec-policy-isakmp-use1-10]ike-peer bca
[Router2-ipsec-policy-isakmp-use1-10]proposal tran1
[Router2-ipsec-policy-isakmp-use1-10]security acl 3001
```

在 Router1 上使用 display ipsec policy 命令查看所配置的信息。

第 11 章　VPN 配置

```
<Router1>display ipsec policy
===========================================
IPSec policy group: "map1"
Using interface:
===========================================
    Sequence number: 10
    Security data flow: 3001
    Peer name    : abc
    Perfect forward secrecy: None
    Proposal name: tran1
    IPSec SA local duration(time based): 3600 seconds
    IPSec SA local duration(traffic based): 1843200 kilobytes
    Anti-replay window size: 32
    SA trigger mode: Automatic
    Route inject: None
    Qos pre-classify: Disable
```

在 Router2 上使用 display ipsec policy 命令查看所配置的信息。

```
<Router2>display ipsec policy
===========================================
IPSec policy group: "use1"
Using interface:
===========================================
    Sequence number: 10
    Security data flow: 3001
    Peer name    : bca
    Perfect forward secrecy: None
    Proposal name: tran1
    IPSec SA local duration(time based): 3600 seconds
    IPSec SA local duration(traffic based): 1843200 kilobytes
    Anti-replay window size: 32
    SA trigger mode: Automatic
    Route inject: None
    Qos pre-classify: Disable
```

结果显示，Router1 和 Roouter2 上的结果与配置的信息一致。

7. 路由器接口应用安全策略组

在路由器 Router1 的接口上使用 ipsec policy 命令引用安全策略组 map1，使接口具有 IPSec 的保护功能。

```
[Router1]interface GigabitEthernet 0/0/1
[Router1-GigabitEthernet0/0/1]ipsec policy map1
```

在路由器 Router2 的接口上引用安全策略组，使接口具有 IPSec 的保护功能。

```
[Router2]interface GigabitEthernet 0/0/1
[Router2-GigabitEthernet0/0/1]ipsec policy use1
```

8. 测试

配置成功后，总公司与分公司的 PC 执行 ping 命令正常，它们之间的数据传输将被加密。在 Router2 上使用 display ipsec statistics esp 命令查看数据包的统计。

```
<Router2>display ipsec statistics esp
  Inpacket count                      : 15
  Inpacket auth count                 : 0
  Inpacket decap count                : 0
  Outpacket count                     : 14
  Outpacket auth count                : 0
  Outpacket encap count               : 0
  Inpacket drop count                 : 0
  Outpacket drop count                : 0
  BadAuthLen count                    : 0
  AuthFail count                      : 0
  InSAAclCheckFail count              : 0
  PktDuplicateDrop count              : 0
  PktSeqNoTooSmallDrop count          : 0
  PktInSAMissDrop count               : 0
```

在 Router1 上使用 display ike sa 命令查看结果。

```
<Router1>display ike sa
  Conn-ID  Peer          VPN  Flag(s)         Phase
  ---------------------------------------------------------
      4    211.85.2.2    0    RD|ST           2
      3    211.85.2.2    0    RD|ST           1
Flag Description:
RD--READY   ST--STAYALIVE  RL--REPLACED  FD--FADING  O--TIMEOUT
HRT--HEARTBEAT  LKG--LAST KNOWN GOOD SEQ NO.  BCK--BACKED UP
```

11.2　Efficient VPN 配置

一、原理概述

IPSec VPN 技术以其高度的安全性、可靠性以及灵活性成为企业构建其 VPN 网络的首选。在包含大量分支的 IPSec 应用中，用户希望各个分支网关的配置能够尽量简单，而复杂的配置集中在总部网关上。Efficient VPN 可以解决 IPSec VPN 分支配置复杂的问题。

Efficient VPN 采用 Client/Server 结构。它将 IPSec 及其他相应配置集中在 Server 端（总部网关），当 Remote 端（分支网关）配置好基本参数时，Remote 端发起协商并与 Server 端建立起 IPSec 隧道，然后 Server 端将 IPSec 的其他相关属性及网络资源"推送"给 Remote 端，简化了分支网关的 IPSec 和其他网络资源的配置和维护。另外，Efficient VPN 还支持远程站点设备的自动升级。

Efficient VPN 运行模式包括以下几种。

（1）Client 模式：一般用于出差员工或小的分支机构通过私网接入总部网络。
（2）Network 模式：一般用于分支和总部 IP 地址已统一规划的网络。
（3）Network-plus 模式：与 Network 模式相比，Remote 端仍会向 Server 端申请 IP 地址。
（4）Network-auto-cfg 模式：与 Network-plus 模式相比，Remote 端还会向 Server 端申请 IP 地址池，IP 地址池用于给用户分配地址。

二、实验目的

1. 理解 Efficient VPN 的基本原理
2. 掌握采用 Network 模式配置 Efficient VPN 的方法

三、实验内容

本实验模拟企业网络环境，Router1 为分公司网关，Router2 为总公司网关，分公司与总公司通过公网建立通信，并且总公司与分公司的网络已统一规划。分公司子网为 192.168.10.0/24，总公司子网为 192.168.20.0/24。需要对分公司子网与总公司子网之间相互访问的流量进行安全保护，并且分公司网关配置能够尽量简单，由总公司网关对分支网关进行集中管理。分公司与总公司通过公网建立通信，可以在 Router1 与 Router2 之间采用 Efficient VPN Network 模式建立一个 IPSec 隧道来实施安全保护，便于 IPSec 隧道的建立与维护管理。

Efficient VPN Network 模式下，Router1 不会向 Router2 申请 IP 地址，而是用原有 IP 地址与 Router2 建立 IPSec 隧道。Router1 向 Router2 申请 DNS 域名、DNS 服务器地址和 WINS 服务器地址，提供给分支子网使用。

四、实验拓扑

Efficient VPN 配置拓扑如图 11.4 所示。

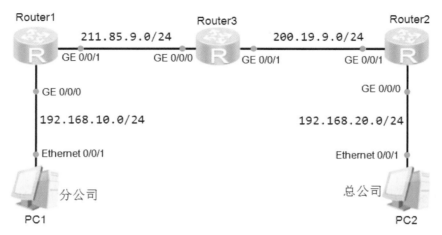

图 11.4　Efficient VPN 配置拓扑

五、实验地址分配

实验地址分配如表 11.2 所示。

表 11.2　实验地址分配

设备	接口	IP 地址	子网掩码	默认网关
Router1（AR2220）	GE 0/0/0	192.168.10.254	255.255.255.0	N/A
	GE 0/0/1	211.85.9.16	255.255.255.0	N/A
Router2（AR2220）	GE 0/0/0	192.168.20.254	255.255.255.0	N/A
	GE 0/0/1	200.19.9.16	255.255.255.0	N/A
Router3（AR2220）	GE 0/0/0	211.85.9.18	255.255.255.0	N/A
	GE 0/0/1	200.19.9.18	255.255.255.0	N/A
PC1	Ethernet 0/0/1	192.168.10.1	255.255.255.0	192.168.10.254
PC2	Ethernet 0/0/1	192.168.20.1	255.255.255.0	192.168.20.254

六、实验步骤

1. 基本配置

1) PC 的基本配置

根据实验地址分配表完成 PC1 的基本配置，如图 11.5 所示。

第 11 章 VPN 配置

图 11.5　PC1 的基本配置

根据实验地址分配表完成 PC2 的基本配置，如图 11.6 所示。

图 11.6　PC2 的基本配置

2) Router1 的基本配置

根据实验地址分配表完成 Router1 的基本配置。

```
<Huawei>system-view
[Huawei]sysname Router1
[Router1]interface GigabitEthernet 0/0/0
[Router1-GigabitEthernet0/0/0]ip address 192.168.10.254 24
[Router1]interface GigabitEthernet 0/0/1
[Router1-GigabitEthernet0/0/1]ip address 211.85.9.16 24
```

3) Router2 的基本配置

根据实验地址分配表完成 Router2 的基本配置。

```
<Huawei>system-view
[Huawei]sysname Router2
[Router2]interface GigabitEthernet 0/0/0
[Router2-GigabitEthernet0/0/0]ip address 192.168.20.254 24
[Router2]interface GigabitEthernet 0/0/1
[Router2-GigabitEthernet0/0/1]ip address 200.19.9.16 24
```

4) Router3 的基本配置

根据实验地址分配表完成 Router3 的基本配置。

```
<Huawei>system-view
[Huawei]sysname Router1
[Router1]interface GigabitEthernet 0/0/0
[Router1-GigabitEthernet0/0/0]ip address 211.85.9.18 24
[Router1]interface GigabitEthernet 0/0/1
[Router1-GigabitEthernet0/0/1]ip address 200.19.9.18 24
```

2. OSPF 网络配置

1) Router1 的 OSPF 配置

```
[Router1]ospf 1
[Router1-ospf-1]area 0
[Router1-ospf-1-area-0.0.0.0]network 192.168.10.0 0.0.0.255
[Router1-ospf-1-area-0.0.0.0]network 211.85.9.0 0.0.0.255
```

2) Router2 的 OSPF 配置

```
[Router2]ospf 1
[Router2-ospf-1]area 0
[Router2-ospf-1-area-0.0.0.0]network 192.168.20.0 0.0.0.255
[Router2-ospf-1-area-0.0.0.0]network 200.19.9.0 0.0.0.255
```

3) Router3 的 OSPF 配置

```
[Internet]ospf 1
[Internet-ospf-1]area 0
[Internet-ospf-1-area-0.0.0.0]network 211.85.9.0 0.0.0.255
[Internet-ospf-1-area-0.0.0.0]network 200.19.9.0 0.0.0.255
```

4) 测试

在 PC1 上使用 ping 命令测试与 PC2 之间的连通性。

```
PC>ping 192.168.20.1
From 192.168.20.1: bytes=32 seq=2 ttl=125 time=31 ms
From 192.168.20.1: bytes=32 seq=2 ttl=125 time=31 ms
From 192.168.20.1: bytes=32 seq=3 ttl=125 time=16 ms
From 192.168.20.1: bytes=32 seq=4 ttl=125 time=31 ms
From 192.168.20.1: bytes=32 seq=5 ttl=125 time=31 ms
```

第 11 章 VPN 配置

```
--- 192.168.20.1 ping statistics ---
 5 packet(s) transmitted
 5 packet(s) received
 20.00% packet loss
 round-trip min/avg/max = 0/27/31 ms
```

结果显示，可以通信，说明 OSPF 网络配置正确。

3. 采用 Network 模式配置 Efficient VPN

1) Router1 的配置

在 Router1 上采用 Network 模式配置 Efficient VPN，作为协商发起方与 Router2 建立 IPSec 隧道。

(1) 配置 ACL。

配置 ACL，定义由子网 192.168.10.0/24 去子网 192.168.20.0/24 的数据流。

```
[Router1]acl number 3001
[Router1-acl-adv-3001]rule 1 permit ip source 192.168.10.0 0.0.0.255
destination 192.168.20.0 0.0.0.255
```

(2) 配置 Efficient VPN。

配置 Efficient VPN 的模式为 Network，并在模式视图下引用 ACL 3001、指定 IKE 协商时的对端 IP 地址 200.19.9.16、预共享密钥 hbeutc123 等。

```
[Router1]ipsec efficient-vpn evpn1 mode network
[Router1-ipsec-efficient-vpn-evpn]security acl 3001
[Router1-ipsec-efficient-vpn-evpn]remote-address 200.19.9.16 v1
[Router1-ipsec-efficient-vpn-evpn]pre-shared-key cipher hbeutc123
[Router1-ipsec-efficient-vpn-evpn]quit
```

(3) 在接口上应用 Efficient VPN。

```
[Router1]interface GigabitEthernet 0/0/1
[Router1-GigabitEthernet1/0/0]ipsec efficient-vpn evpn1
```

2) Router2 的配置

在 Router2 上配置策略模板方式的安全策略，作为协商响应方与 Router1 建立 IPSec 隧道。

(1) 通过 AAA 业务模板配置要推送的资源属性。

```
[Router2]aaa
[Router2-aaa]service-scheme schemetest
[Router2-aaa-service-schemetest]dns 2.2.2.2
[Router2-aaa-service-schemetest]dns 2.2.2.3 secondary
[Router2-aaa-service-schemetest]quit
```

(2) 配置 IKE 安全提议和 IKE 对等体。

```
[Router2]ike proposal 5
[Router2-ike-proposal-5]dh group2
[Router2-ike-proposal-5]encryption-algorithm 3des-cbc
[Router2-ike-proposal-5]quit
[Router2]ike peer peer123 v1
```

(3) 配置 IPSec 安全提议、策略模板和策略组。

```
[Router2]ipsec proposal tran1
[Router2-ipsec-proposal-tran1]quit
[Router2]ipsec policy-template user1 10
[Router2-ipsec-policy-templet-use1-10]ike-peer peer123
[Router2-ipsec-policy-templet-use1-10]proposal tran1
[Router2-ipsec-policy-templet-use1-10]quit
[Router2]ipsec policy policy1 10 isakmp template user1
```

(4) 在接口上应用安全策略组。

```
[Router2]interface GigabitEthernet 0/0/1
[Router2-GigabitEthernet1/0/0]ipsec policy policy1
[Router2-GigabitEthernet1/0/0]quit
```

3) 测试

配置成功后，在主机 PC1 上执行 ping 操作仍然可以 ping 通主机 PC2。

(1) 测试 IKE 协商的信息。

在 Router1 上执行 display ike sa 查看由 IKE 协商建立的安全联盟信息。

```
[Router1] display ike sa
  Conn-ID   Peer            VPN     Flag(s)         Phase
------------------------------------------------------------
    210    211.85.9.16       0       RD              2
```

第 11 章 VPN 配置

```
   201    211.85.9.16     0      RD                    1
Flag Description:
RD--READY   ST--STAYALIVE   RL--REPLACED  FD--FADING  TO--TIMEOUT
HRT--HEARTBEAT  LKG--LAST  KNOWN  GOOD  SEQ  NO.  BCK--BACKED UP
```

在 Router2 上执行 **display ike sa** 查看由 IKE 协商建立的安全联盟信息。

```
[Router2] display ike sa
  Conn-ID  Peer          VPN   Flag(s)          Phase
---------------------------------------------------------------
   61     211.85.9.16    0     RD                2
   58     211.85.9.16    0     RD                1
Flag Description:
RD--READY   ST--STAYALIVE   RL--REPLACED  FD--FADING  TO--TIMEOUT
HRT--HEARTBEAT  LKG--LAST  KNOWN  GOOD  SEQ  NO.  BCK--BACKED UP
```

结果显示,两端设备的 IKE SA 已经建立成功。

在路由器 Router2 上使用 **display ipsec sa** 命令查看 IPsec 安全联盟的配置信息。

```
<Router2>display ipsec sa
===============================
Interface: GigabitEthernet0/0/1
Path MTU: 1500
===============================
  -----------------------------
  IPSec policy name: "policy1"
  Sequence number  : 10
  Acl Group        : 0
  Acl rule         : 0
  Mode             : Template
  -----------------------------
    Connection ID     : 61
    Encapsulation mode: Tunnel
    Tunnel local      : 200.19.9.16
    Tunnel remote     : 211.85.9.16
    Flow source       : 192.168.20.0/255.255.255.0 0/0
    Flow destination  : 192.168.10.0/255.255.255.0 0/0
    Qos pre-classify  : Disable
    [Outbound ESP SAs]
      SPI: 3538523129 (0xd2e993f9)
      Proposal: ESP-ENCRYPT-AES-128 ESP-AUTH-SHA1
      SA remaining key duration (bytes/sec): 1887436800/1311
      Max sent sequence-number: 0
      UDP encapsulation used for NAT traversal: N
    [Inbound ESP SAs]
```

```
SPI: 2185958894 (0x824b15ee)
Proposal: ESP-ENCRYPT-AES-128 ESP-AUTH-SHA1
SA remaining key duration (bytes/sec): 1887436800/1311
Max received sequence-number: 0
Anti-replay window size: 32
UDP encapsulation used for NAT traversal: N
```

结果显示，隧道的本地 IP 地址、远程 IP 地址，两端私有网络信息、加密信息等。

11.3 GRE VPN 协议基本配置

一、原理概述

通用路由封装协议（generic routing encapsulation，GRE）可以对某些网络层协议（如 IPX、ATM、IPv6 等）的数据报文进行封装，使这些被封装的数据报文能够在另一个网络层协议（如 IPv4）中传输。GRE 提供了将一种协议的报文封装在另一种协议报文中的机制，是一种三层隧道封装技术，使报文可以通过 GRE 隧道透明传输，解决异种网络的传输问题。

Tunnel 是一个虚拟的点对点的连接，提供了一条通路使封装的数据报文能够在这个通路上传输，并且在一个 Tunnel 的两端分别对数据报进行封装及解封装。一个 X 协议的报文要想穿越 IP 网络在 Tunnel 中传输，必须要经过加封装与解封装两个过程。

GRE 协议的特点如下：

(1) GRE 实现机制简单，对隧道两端的设备负担小。

(2) GRE 隧道可以通过 IPv4 网络连通多种网络协议的本地网络，有效利用了原有的网络架构，降低成本。

(3) GRE 隧道扩展了跳数受限网络协议的工作范围，支持企业灵活设计网络拓扑。

(4) GRE 隧道可以封装组播数据，结合 IPSec 可以保证语音、视频等组播业务的安全。

(5) GRE 隧道支持使能 MPLS LDP，使用 GRE 隧道承载 MPLS LDP 报文，建立 LDP LSP，实现 MPLS 骨干网的互通。

(6) GRE 隧道将不连续的子网连接起来，组建 VPN 实现企业总部和分支间安全的连接。

二、实验目的

1. 理解 GRE 的基本原理
2. 掌握配置 GRE 隧道的方法
3. 掌握配置基于 GRE 接口的动态路由协议的方法

三、实验内容

本实验模拟企业网络环境,总公司的主机 PC1 和分公司的主机 PC2 分别通过路由器 Router1 和 Router2 接入模拟 Internet 的路由器 Router3。要求 Router1、Router2、Router3 使用 OSPF 协议路由实现公网互通。在 PC1 和 PC2 上运行 IPv4 私网协议,现需要 PC1 和 PC2 通过公网实现 IPv4 私网互通。

四、实验拓扑

GRE 配置拓扑如图 11.7 所示。

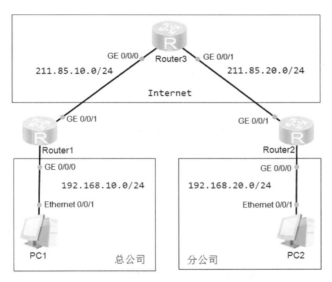

图 11.7 GRE 配置拓扑

五、实验地址分配

实验地址分配如表 11.3 所示。

表 11.3 实验地址分配

设备	接口	IP 地址	子网掩码	默认网关
Router1 (AR2220)	GE 0/0/0	192.168.10.1	255.255.255.0	N/A
	GE 0/0/1	211.85.10.1	255.255.255.0	
Router2 (AR2220)	GE 0/0/0	192.168.20.1	255.255.255.0	N/A
	GE 0/0/1	211.85.20.2	255.255.255.0	
Router3 (AR2220)	GE 0/0/0	211.85.10.2	255.255.255.0	
	GE 0/0/1	211.85.20.1	255.255.255.0	

设备	接口	IP 地址	子网掩码	默认网关
PC1	Ethernet 0/0/1	192.168.10.2	255.255.255.0	192.168.10.1
PC2	Ethernet 0/0/1	192.168.20.2	255.255.255.0	192.168.20.1

六、实验步骤

1. 基本配置

1) Router1 的基本配置

根据实验地址分配表完成 Router1 的基本配置。

```
<Huawei>system-view
[Huawei]sysname Router1
[Router1]interface GigabitEthernet 0/0/0
[Router1-GigabitEthernet0/0/0]ip address 192.168.10.1 24
[Router1-GigabitEthernet0/0/0]interface GigabitEthernet 0/0/1
[Router1-GigabitEthernet0/0/1]ip address 211.85.10.1 24
```

2) Router2 的基本配置

根据实验地址分配表完成 Router2 的基本配置。

```
<Huawei>system-view
[Huawei]sysname Router2
[Router2]interface GigabitEthernet 0/0/0
[Router2-GigabitEthernet0/0/0]ip address 192.168.20.1 24
[Router2-GigabitEthernet0/0/0]interface GigabitEthernet 0/0/1
[Router2-GigabitEthernet0/0/1]ip address 211.85.20.2 24
```

3) Router3 的基本配置

根据实验地址分配表完成 Router3 的基本配置。

```
<Huawei>system-view
[Huawei]sysname Router3
[Router3]interface GigabitEthernet 0/0/0
[Router3-GigabitEthernet0/0/0]ip address 211.85.10.2 24
[Router3-GigabitEthernet0/0/0]interface GigabitEthernet 0/0/1
[Router3-GigabitEthernet0/0/1]ip address 211.85.20.1 24
```

4) PC1 的基本配置

根据实验地址分配表完成 PC1 的基本配置，如图 11.8 所示。

图 11.8 PC1 的基本配置

5) PC2 的基本配置

根据实验地址分配表完成 PC2 的基本配置，如图 11.9 所示。

图 11.9 PC2 的基本配置

2. 骨干网上配置 OSPF

1) Router1 的 OSPF 配置

```
[Router1]ospf 1
[Router1-ospf-1]area 0
[Router1-ospf-1-area-0.0.0.0]network 211.85.10.0 0.0.0.255
```

2) Router2 的的 OSPF 配置

```
[Router2]ospf 1
[Router2-ospf-1]area 0
[Router2-ospf-1-area-0.0.0.0]network 211.85.20.0 0.0.0.255
```

3) Router3 的的 OSPF 配置

```
[Router3]ospf 1
[Router3-ospf-1]area 0
```

```
[Router3-ospf-1-area-0.0.0.0]network 211.85.10.0 0.0.0.255
[Router3-ospf-1-area-0.0.0.0]network 211.85.20.0 0.0.0.255
```

4) 查看 Router1 的路由表

```
[Router1]display ip routing-table protocol ospf
Route Flags: R - relay, D - download to fib
------------------------------------------------------------------
Public routing table : OSPF
        Destinations : 1        Routes : 1
OSPF routing table status : <Active>
        Destinations : 1        Routes : 1
Destination/Mask  Proto  Pre  Cost  Flags  NextHop       Interface
  211.85.20.0/24   OSPF   10   2     D     211.85.10.2 GigabitEthernet0/0/1
OSPF routing table status : <Inactive>
        Destinations : 0        Routes : 0
```

结果显示，Router1 学习到去往 211.85.20.0 网络的路由。

5) 查看 Router2 的路由表

```
[Router2]display ip routing-table protocol ospf
Route Flags: R - relay, D - download to fib
------------------------------------------------------------------
Public routing table : OSPF
        Destinations : 1        Routes : 1
OSPF routing table status : <Active>
        Destinations : 1        Routes : 1
Destination/Mask  Proto  Pre  Cost  Flags  NextHop       Interface
  211.85.10.0/24   OSPF   10   2     D     211.85.20.1 GigabitEthernet0/0/1
OSPF routing table status : <Inactive>
        Destinations : 0        Routes : 0
```

结果显示，Router2 学习到去往 211.85.10.0 网络的路由。

6) 在分公司 PC2 上测试与总公司 PC1 之间的连通性

```
PC>ping 192.168.10.2
ping 192.168.10.2: 32 data bytes, Press Ctrl_C to break
Request timeout!
Request timeout!
Request timeout!
Request timeout!
Request timeout!
--- 192.168.10.2 ping statistics ---
```

```
  5 packet(s) transmitted
  0 packet(s) received
  100.00% packet loss
```

结果显示，跨越了 Internet 的两个私有网络之间默认是无法直接通信的，可以通过 GRE 来实现通信。

3. 配置 GRE Tunnel

在路由器 Router1 和 Router2 上配置 GRE Tunnel，使用 **interface tunnel** 命令创建隧道接口，指定隧道模式为 GRE。配置 Router1 Tunnel 接口的源地址为 GE 0/0/1 接口的 IP 地址，目的地址为 Router2 的 GE 0/0/1 接口的 IP 地址。使用 **ip address** 命令配置 Tunnel 接口的 IP 地址，注意要在同一个网段。

1) Router1 的 Tunnel 接口配置

```
[Router1]interface tunnel 0/0/0
[Router1-Tunnel0/0/0]tunnel-protocol gre
[Router1-Tunnel0/0/0]source 211.85.10.1
[Router1-Tunnel0/0/0]destination 211.85.20.2
[Router1-Tunnel0/0/0]ip address 192.168.0.1 24
```

2) Router2 的 Tunnel 接口配置

```
[Router2]interface tunnel 0/0/0
[Router2-Tunnel0/0/0]tunnel-protocol gre
[Router2-Tunnel0/0/0]source 211.85.20.2
[Router2-Tunnel0/0/0]destination 211.85.10.1
[Router2-Tunnel0/0/0]ip address 192.168.0.2 24
```

3) 测试隧道的连通性

在 Router1 上使用 ping 命令测试本端隧道接口地址与目的隧道接口地址的连通性。

```
[Router1]ping -a 192.168.0.1 192.168.0.2
  ping 192.168.0.2: 56  data bytes, press CTRL_C to break
    Reply from 192.168.0.2: bytes=56 Sequence=1 ttl=255 time=260 ms
    Reply from 192.168.0.2: bytes=56 Sequence=2 ttl=255 time=30 ms
    Reply from 192.168.0.2: bytes=56 Sequence=3 ttl=255 time=30 ms
    Reply from 192.168.0.2: bytes=56 Sequence=4 ttl=255 time=20 ms
    Reply from 192.168.0.2: bytes=56 Sequence=5 ttl=255 time=20 ms
  --- 192.168.0.2 ping statistics ---
    5 packet(s) transmitted
    5 packet(s) received
```

计算机网络实践教程

```
        0.00% packet loss
        round-trip min/avg/max = 20/72/260 ms
```

结果显示,通信正常。

4) 查看隧道接口状态

在 Router1 上使用 **display interface tunnel** 命令查看隧道接口状态。

```
    [Router1]display interface tunnel 0/0/0
    Tunnel0/0/0 current state : UP
    Line protocol current state : UP
    Last line protocol up time : 2022-10-01 17:11:56 UTC-08:00
    Description:211.85.20.2
    Route Port,The Maximum Transmit Unit is 1500
    Internet Address is 192.168.0.1/24
    Encapsulation is TUNNEL, loopback not set
    Tunnel source 211.85.10.1 (GigabitEthernet0/0/1), destination 211.85.20.2
    Tunnel protocol/transport GRE/IP, key disabled
    keepalive disabled
    ……
```

在 Router2 上使用 display interface tunnel 命令查看隧道接口状态。

```
    [Router2]display interface tunnel 0/0/0
    Tunnel0/0/0 current state : UP
    Line protocol current state : UP
    Last line protocol up time : 2022-10-01 17:12:41 UTC-08:00
    Description:211.85.10.1
    Route Port,The Maximum Transmit Unit is 1500
    Internet Address is 192.168.0.2/24
    Encapsulation is TUNNEL, loopback not set
    Tunnel source 211.85.20.2 (GigabitEthernet0/0/1),destination 211.85.10.1
    ……
```

结果显示了当前隧道两端的状态为正常启动状态,链路协议状态为运行状态,隧道封装协议为 GRE,Tunnel 的 IP 地址及配置的隧道源地址和目的地址分别为 Router1 和 Router2 的 GE 0/0/1 接口 IP 地址。

4. 配置基于 GRE 接口的动态路由协议

1) 查看 Router1 和 Router2 的路由表

在 Router1 上使用 **display ip routing-table** 命令查看路由表。

```
[Router1]display ip routing-table
Route Flags: R - relay, D - download to fib
------------------------------------------------------------------------
Routing Tables: Public
         Destinations : 14       Routes : 14
Destination/Mask    Proto  Pre  Cost   Flags NextHop         Interface
     192.168.0.0/24  Direct  0    0      D    192.168.0.1    Tunnel0/0/0
     192.168.0.1/32  Direct  0    0      D    127.0.0.1      Tunnel0/0/0
     192.168.10.0/24 Direct  0    0      D    192.168.10.1   GigabitEthernet0/0/0
     192.168.10.1/32 Direct  0    0      D    127.0.0.1      GigabitEthernet0/0/0
     211.85.10.0/24  Direct  0    0      D    211.85.10.1    GigabitEthernet0/0/1
     211.85.10.1/32  Direct  0    0      D    127.0.0.1      GigabitEthernet0/0/1
     211.85.20.0/24  OSPF    10   2      D    211.85.10.2    GigabitEthernet0/0/1
     ……
```

在 Router2 上使用 display ip routing-table 命令查看路由表。

```
[Router2]display ip routing-table
Route Flags: R - relay, D - download to fib
------------------------------------------------------------------------
Routing Tables: Public
         Destinations : 10       Routes : 10
Destination/Mask    Proto  Pre  Cost   Flags NextHop         Interface
     192.168.20.0/24 Direct  0    0      D    192.168.20.1   GigabitEthernet0/0/0
     192.168.20.1/32 Direct  0    0      D    127.0.0.1      GigabitEthernet0/0/0
     211.85.20.0/24  Direct  0    0      D    211.85.20.2    GigabitEthernet0/0/1
     211.85.20.2/32  Direct  0    0      D    127.0.0.1      GigabitEthernet0/0/1
     ……
```

结果显示，Router1 和 Router2 的路由表中已经配置隧道接口的路由，但没有对方私有网络路由信息。

2) 配置静态路由表

分别在 Router1 和 Router2 上使用 ip route-static 命令配置静态路由。

```
[Router1]ip route-static 192.168.20.0 255.255.255.0 tunnel 0/0/0

[Router2]ip route-static 192.168.10.0 255.255.255.0 tunnel 0/0/0
```

或配置 RIP 动态路由表，二选一，不能重复配置路由表。
分别在 Router1 和 Router2 上使用 RIP 协议配置动态路由。

```
[Router1]rip 1
[Router1-rip-1]version 2
[Router1-rip-1]network 192.168.10.0
[Router1-rip-1]network 192.168.0.0

[Router2]rip 1
[Router2-rip-1]version 2
[Router2-rip-1]network 192.168.20.0
[Router2-rip-1]network 192.168.0.0
```

3) 查看路由表

在 Router1 上使用 display ip routing-table 命令查看路由表。

```
[Router1]display ip routing-table
Route Flags: R - relay, D - download to fib
------------------------------------------------------------------------
Routing Tables: Public
        Destinations : 15      Routes : 15
Destination/Mask    Proto   Pre  Cost  Flags  NextHop        Interface
 ......
 192.168.0.0/24     Direct  0    0       D    192.168.0.1    Tunnel0/0/0
 192.168.0.1/32     Direct  0    0       D    127.0.0.1      Tunnel0/0/0
 192.168.0.255/32   Direct  0    0       D    127.0.0.1      Tunnel0/0/0
 ......
```

在 Router2 上使用 display ip routing-table 命令查看路由表。

```
  [Router2]display ip routing-table
Route Flags: R - relay, D - download to fib
------------------------------------------------------------------------
Routing Tables: Public
        Destinations : 15      Routes : 15
Destination/Mask    Proto   Pre  Cost  Flags  NextHop        Interface
 ......
```

```
192.168.0.0/24       Direct 0    0         D    192.168.0.2    Tunnel0/0/0
192.168.0.2/32       Direct 0    0         D    127.0.0.1      Tunnel0/0/0
192.168.0.255/32     Direct 0    0         D    127.0.0.1      Tunnel0/0/0
......
```

结果显示，在 Router1 和 Router2 上添加了对方私有网络的路由信息。

4) 查看 Router3 的路由表

```
[Router3]display ip routing-table
Route Flags: R - relay, D - download to fib
------------------------------------------------------------------
Routing Tables: Public
        Destinations : 10    Routes : 10
Destination/Mask       Proto  Pre Cost Flags NextHop        Interface
    127.0.0.0/8        Direct 0    0         D    127.0.0.1      InLoopBack0
    127.0.0.1/32       Direct 0    0         D    127.0.0.1      InLoopBack0
127.255.255.255/32     Direct 0    0         D    127.0.0.1      InLoopBack0
    211.85.10.0/24     Direct 0    0         D    211.85.10.2    GigabitEthernet0/0/0
    211.85.10.2/32     Direct 0    0         D    127.0.0.1      GigabitEthernet0/0/0
  211.85.10.255/32     Direct 0    0         D    127.0.0.1      GigabitEthernet0/0/0
    211.85.20.0/24     Direct 0    0         D    211.85.20.1    GigabitEthernet0/0/1
    211.85.20.1/32     Direct 0    0         D    127.0.0.1      GigabitEthernet0/0/1
  211.85.20.255/32     Direct 0    0         D    127.0.0.1      GigabitEthernet0/0/1
255.255.255.255/32     Direct 0    0         D    127.0.0.1      InLoopBack0
```

结果显示，路由器 Router3 上没有任何 Router1 和 Router2 的各自私有网段的路由信息。说明通过 GRE 建立起来的隧道，能够跨越公网传递各个内部私有网络的路由信息，实现了两个私有网络间的跨公网通信。

5) 测试分公司 PC 与总公司 PC 之间的连通性

在分公司 PC2 上测试与总公司 PC1 之间的连通性。

```
PC>ping 192.168.10.2
ping 192.168.10.2: 32 data bytes, Press Ctrl_C to break
From 192.168.10.2: bytes=32 seq=1 ttl=126 time=31 ms
From 192.168.10.2: bytes=32 seq=2 ttl=126 time=32 ms
From 192.168.10.2: bytes=32 seq=3 ttl=126 time=31 ms
From 192.168.10.2: bytes=32 seq=4 ttl=126 time=31 ms
From 192.168.10.2: bytes=32 seq=5 ttl=126 time=31 ms
--- 192.168.10.2 ping statistics ---
 5 packet(s) transmitted
 5 packet(s) received
 0.00% packet loss
 round-trip min/avg/max = 31/31/32 ms
```

结果显示,通信正常。

11.4 MPLS/LDP 基本配置

一、原理概述

多协议标签交换(multi-protocol label switching,MPLS)是一种 IP 骨干网技术,将第三层路由技术和二层交换技术结合,充分发挥了 IP 路由的灵活性和二层交换的便捷性。MPLS 并不是一种业务或应用,而是一种隧道技术。这种技术不仅支持多种高层协议与业务,而且在一定程度上可以保证信息传输的安全性。

MPLS 网络是指由运行 MPLS 协议的交换节点构成的区域。这些交换节点就是 MPLS 标记交换路由器,按照它们在 MPLS 网络中所处位置的不同,可划分为 MPLS 标记边缘路由器(label edge router,LER)和 MPLS 标记核心路由器(label switching router,LSR)。LER 位于 MPLS 网络边缘与其他网络或者用户相连;LSR 位于 MPLS 网络内部。两类路由器的功能因其在网络中位置的不同而略有差异。

MPLS 作为一种分类转发技术,将具有相同转发处理方式的分组归为一类,称为转发等价类(forwarding equivalence class,FEC)。相同 FEC 的分组在 MPLS 网络中获得完全相同的处理。FEC 的划分方式非常灵活,可以是以源地址、目的地址、源端口、目的端口、协议类型或 VPN 等为划分依据的任意组合。例如,在传统的采用最长匹配算法的 IP 转发中,到同一个目的地址的所有报文就是一个 FEC。

MPLS 的工作过程如下:

(1) LDP(label distribution protocol)和传统路由协议(RIP、OSPF 等)一起,在 LSR 中为有需求的 FEC 建立路由表和标签映射表。

(2) 入节点接收分组,完成第三层功能,判定分组所属的 FEC,并给分组加上标签,形成 MPLS 标签分组,转发到中间节点 Transit。

(3) Transit 根据分组上的标签以及标签转发表进行转发,不对标签分组进行第三层处理。

(4) 在出节点去掉分组中的标签,继续进行后面的转发。

二、实验目的

1. 掌握 MPLS 和 LDP 的工作原理
2. 掌握 MPLS 标签转发的过程

三、实验内容

本实验模拟企业网络环境，全网运行 OSPF，需要部署 MPLS 和 LDP，在分公司路由器 Router1 与总部路由器 Router3 之间创建 LSP。

四、实验拓扑

MPLS/LDP 基本配置拓扑如图 11.10 所示。

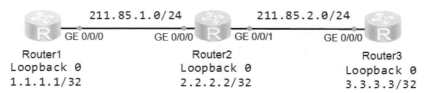

图 11.10　MPLS/LDP 基本配置拓扑

五、实验地址分配

实验地址分配如表 11.4 所示。

表 11.4　实验地址分配

设备	接口	IP 地址	子网掩码	默认网关
Router1 （AR2220）	GE 0/0/0	211.85.1.1	255.255.255.0	N/A
	Loopback 0	1.1.1.1	255.255.255.255	N/A
Router2 （AR2220）	GE 0/0/0	211.85.1.2	255.255.255.0	N/A
	GE 0/0/1	211.85.2.1	255.255.255.0	N/A
	Loopback 0	2.2.2.2	255.255.255.255	N/A
Router3 （AR2220）	GE 0/0/0	211.85.2.2	255.255.255.0	N/A
	Loopback 0	3.3.3.3	255.255.255.255	N/A

六、实验步骤

1. 基本配置

1）Router1 的基本配置

根据实验地址分配表完成 Router1 的基本配置。

```
<Huawei>system-view
[Huawei]sysname Router1
[Router1]interface GigabitEthernet 0/0/0
```

```
[Router1-GigabitEthernet0/0/0]ip address 211.85.1.1 24
[Router1-GigabitEthernet0/0/0]interface loopback 0
[Router1-LoopBack0]ip address 1.1.1.1 32
[Router1-LoopBack0]
```

2) Router2 的基本配置

根据实验地址分配表完成 Router2 的基本配置。

```
<Huawei>system-view
[Huawei]sysname Router2
[Router2]interface GigabitEthernet 0/0/0
[Router2-GigabitEthernet0/0/0]ip address 211.85.1.2 24
[Router2-GigabitEthernet0/0/0]interface GigabitEthernet 0/0/1
[Router2-GigabitEthernet0/0/1]ip address 211.85.2.1 24
[Router2-GigabitEthernet0/0/1]interface loopback 0
[Router2-LoopBack0]ip address 2.2.2.2 32
[Router2-LoopBack0]
```

3) Router3 的基本配置

根据实验地址分配表完成 Router3 的基本配置。

```
<Huawei>system-view
[Huawei]sysname Router3
[Router3]interface GigabitEthernet 0/0/0
[Router3-GigabitEthernet0/0/0]ip address 211.85.2.2 24
[Router3-GigabitEthernet0/0/0]interface loopback 0
[Router3-LoopBack0]ip address 3.3.3.3 32
[Router3-LoopBack0]
```

2. OSPF 配置

OSPF 配置中，需要使用 Router-ID 作为路由器的身份。Router-ID 选举规则为，可以使用 Router-ID 命令进行配置，在未配置 Router-ID 时，首先选择 Loopback 接口地址中最大的地址；如果没有使用 Loopback 接口，则从其他接口的 IP 地址中选择最大的地址作为 Router-ID。

```
[Router1]ospf router-id 1.1.1.1
[Router1-ospf-1]area 0
[Router1-ospf-1-area-0.0.0.0]network 211.85.1.0 0.0.0.255
[Router1-ospf-1-area-0.0.0.0]network 1.1.1.1 0.0.0.0

[Router2]ospf router-id 2.2.2.2
[Router2-ospf-1]area 0
```

```
[Router2-ospf-1-area-0.0.0.0]network 211.85.1.0 0.0.0.255
[Router2-ospf-1-area-0.0.0.0]network 211.85.2.0 0.0.0.255
[Router2-ospf-1-area-0.0.0.0]network 2.2.2.2 0.0.0.0

[Router3]ospf router-id 3.3.3.3
[Router3-ospf-1]area 0
[Router3-ospf-1-area-0.0.0.0]network 211.85.2.0 0.0.0.255
[Router3-ospf-1-area-0.0.0.0]network 3.3.3.3 0.0.0.0
```

在 Router1 上使用 display ospf routing 命令查看路由表。

```
[Router1]display ospf routing
 OSPF Process 1 with Router ID 1.1.1.1
                  Routing Tables
 Routing for Network
 Destination       Cost    Type       NextHop         AdvRouter       Area
 1.1.1.1/32        0       Stub       1.1.1.1         1.1.1.1         0.0.0.0
 211.85.1.0/24     1       Transit    211.85.1.       1.1.1.1         0.0.0.0
 2.2.2.2/32        1       Stub       211.85.1.2      2.2.2.2         0.0.0.0
 3.3.3.3/32        2       Stub       211.85.1.2      3.3.3.3         0.0.0.0
 211.85.2.0/24     2       Transit    211.85.1.2      2.2.2.2         0.0.0.0
 Total Nets: 5
 Intra Area: 5  Inter Area: 0  ASE: 0  NSSA: 0
```

结果显示，路由器 Router1 路由表中存在到达各个不同的网络的路由。OSPF 正常工作。
在 Router1 上使用 ping 命令测试与 Router3 之间的连通性。

```
<Router1>ping 211.85.2.2
 ping 211.85.2.2: 56  data bytes, press CTRL_C to break
   Reply from 211.85.2.2: bytes=56 Sequence=1 ttl=254 time=50 ms
   Reply from 211.85.2.2: bytes=56 Sequence=2 ttl=254 time=30 ms
   Reply from 211.85.2.2: bytes=56 Sequence=3 ttl=254 time=20 ms
   Reply from 211.85.2.2: bytes=56 Sequence=4 ttl=254 time=10 ms
   Reply from 211.85.2.2: bytes=56 Sequence=5 ttl=254 time=30 ms
 --- 211.85.2.2 ping statistics ---
   5 packet(s) transmitted
   5 packet(s) received
   0.00% packet loss
   round-trip min/avg/max = 10/28/50 ms
```

结果显示，远程路由器 Router1 与 Router3 之间可以通信。

3. 配置 MPLS 协议

配置 MPLS 时，需要先配置 LSR ID。在 Router1 上，使用 **mpls lsr-id** 命令配置 Router1 的

LSR ID 为 1.1.1.1。然后使用 mpls 命令全局启用 MPLS，在转发 MPLS 报文的接口 GE 0/0/0 上使用 mpls 命令启动该接口的 MPLS 功能，最后使用 display mpls lsp 命令查看 LSP 信息。

```
[Router1]mpls lsr-id 1.1.1.1
[Router1]mpls
Info: Mpls starting, please wait... OK!
[Router1-mpls]interface GigabitEthernet 0/0/0
[Router1-GigabitEthernet0/0/0]mpls
[Router1-GigabitEthernet0/0/0]quit
[Router1]display mpls lsp
[Router1]                        //LSP为空
```

结果显示，Router1 还不存在任何 LSP。

4. 配置静态 LSP

在 Router1 上配置从 Router1 到 Router3 的静态 LSP 的 Ingress，并进行标签的分配。

```
[Router1]static-lsp ingress 1to3 destination 3.3.3.3 32 nexthop 211.85.1.2 out-label 102
```

在 Router2 上配置从 Router1 到 Router3 的静态 LSP 的 Transit，并进行标签的分配。

```
[Router2]mpls lsr-id 2.2.2.2
[Router2]mpls
[Router2-mpls]interface GigabitEthernet 0/0/0
[Router2-GigabitEthernet0/0/0]mpls
[Router2-GigabitEthernet0/0/0]interface GigabitEthernet 0/0/1
[Router2-GigabitEthernet0/0/1]mpls
[Router2-GigabitEthernet0/0/1]quit
[Router2]static-lsp transit 1to3 incoming-interface GigabitEthernet 0/0/0 in-label 102 nexthop 211.85.2.2 out-label 203
```

在 Router3 上配置从 Router1 到 Router3 的静态 LSP 的 Egress，并进行标签的分配。

```
[Router3]mpls lsr-id 3.3.3.3
[Router3]mpls
[Router3-mpls]interface GigabitEthernet 0/0/0
[Router3-GigabitEthernet0/0/0]mpls
[Router3-GigabitEthernet0/0/0]quit
[Router3]static-lsp egress 1to3 incoming-interface GigabitEthernet 0/0/0 in-lab 203
```

第 11 章 VPN 配置

配置完成后，在 Router1 上使用 display mpls lsp 命令查看 LSP 信息。

```
<Router1>display mpls lsp
------------------------------------------------------------
              LSP Information: STATIC LSP
------------------------------------------------------------
FEC            In/Out Label    In/Out IF      Vrf Name
3.3.3.3/32     NULL/102        -/GE0/0/0
```

结果显示，Router1 上已经拥有了去往 Router3（3.3.3.3/32）的静态 LSP，且在本地的 In 标签为 NULL，说明 Router1 是该 LSP 的 Ingress。

在 Router2 和 Router3 上显示类似的信息。

```
<Router2>display mpls lsp
------------------------------------------------------------
              LSP Information: STATIC LSP
------------------------------------------------------------
FEC            In/Out Label    In/Out IF      Vrf Name
-/-            102/203         GE0/0/0/GE0/0/1

<Router3>display mpls lsp
------------------------------------------------------------
              LSP Information: STATIC LSP
------------------------------------------------------------
FEC            In/Out Label    In/Out IF      Vrf Name
-/-            203/NULL        GE0/0/0/-
```

结果显示，正确配置了从 Router1 到 Router3 的 LSP 信息。

在 Router1 上使用 tracert lsp ip 命令验证去往 3.3.3.3/32 的 MPLS 报文所经过的路径，进一步验证 LSP 的正确性。

```
<Router1>tracert lsp ip 3.3.3.3 32
  LSP Trace Route FEC: IPV4 PREFIX 3.3.3.3/32 , press CTRL_C to break.
  TTL    Replier           Time      Type      Downstream
  0                                  Ingress   211.85.1.2/[102 ]
  1      211.85.1.2        30 ms     Transit   211.85.2.2/[203 ]
  2      3.3.3.3           20 ms     Egress
```

结果显示，MPLS 报文转发过程中使用的标签，以及各路由器的角色信息正确。

在 Router3 上使用 tracert lsp ip 命令验证 LSP 去往 1.1.1.1/32 的 MPLS 报文所经过的路径。

253

```
<Router3>tracert lsp ip 1.1.1.1 32
Error: The specified LSP does not exist.
```

结果显示，LSP 并不存在，因为 LSP 是单向的。

依次在 Router3、Router2 和 Router1 上配置从 Router3 到 Router1 的静态 LSP。

```
[Router3]static-lsp ingress 3to1 destination 1.1.1.1 32 nexthop 211.85.2.1 out-label 302

[Router2]static-lsp transit 3to1 incoming-interface GigabitEthernet 0/0/1 in-label 302 nexthop 211.85.1.1 out-label 201

[Router1]static-lsp egress 3to1 incoming-interface GigabitEthernet 0/0/0 in-label 201
```

配置完成后，在 Router3 上使用 tracert lsp ip 命令验证到 1.1.1.1/32 的 MPLS 报文经过的路径。

```
<Router3>tracert lsp ip 1.1.1.1 32
  LSP Trace Route FEC: IPV4 PREFIX 1.1.1.1/32 , press CTRL_C to break.
  TTL    Replier         Time      Type       Downstream
  0                                Ingress    211.85.2.1/[302 ]
  1      211.85.2.1      30 ms     Transit    211.85.1.1/[201 ]
  2      1.1.1.1         60 ms     Egress
```

结果显示，从 Router3 到 Router1 方向的 LSP 信息完整。

5. 配置动态 LSP

首先，在 Router1、Router2 和 Router3 上删除静态 LSP。

```
[Router1]undo static-lsp ingress 1to3
[Router1]undo static-lsp egress 3to1

[Router2]undo static-lsp transit 1to3
[Router2]undo static-lsp transit 3to1

[Router3]undo static-lsp egress 1to3
[Router3]undo static-lsp ingress 3to1
```

在 Router1 上使用 **mpls ldp** 命令全局启用 LDP，在接口 GE 0/0/0 上使用同样的命令启用 LDP。

```
[Router1]mpls ldp
[Router1-mpls-ldp]interface GigabitEthernet 0/0/0
[Router1-GigabitEthernet0/0/0]mpls ldp
```

在 Router2 上使用 mpls ldp 命令全局启用 LDP，在接口 GE 0/0/0 和 GE 0/0/1 上启用 LDP。

```
[Router2]mpls ldp
[Router2-mpls-ldp]interface GigabitEthernet 0/0/0
[Router2-GigabitEthernet0/0/0]mpls ldp
[Router2-GigabitEthernet0/0/0]interface GigabitEthernet 0/0/1
[Router2-GigabitEthernet0/0/1]mpls ldp
```

在 Router3 上使用 mpls ldp 命令全局启用 LDP，在接口 GE 0/0/0 上启用 LDP。

```
[Router3]mpls ldp
[Router3-mpls-ldp]interface GigabitEthernet 0/0/0
[Router3-GigabitEthernet0/0/0]mpls ldp
```

在 Router1 上使用 display mpls ldp interface 命令查看启用了 LDP 的接口。

```
<Router1>display mpls ldp interface
 LDP Interface Information in Public Network
 Codes:LAM(Label Advertisement Mode), IFName(Interface name)
 A '*' before an interface means the entity is being deleted.
------------------------------------------------------------------
 IFName      Status      LAM    TransportAddress    HelloSent/Rcv
------------------------------------------------------------------
 GE0/0/0     Active      DU     1.1.1.1             122/100
------------------------------------------------------------------
```

结果显示，在 Router1 的接口 GE 0/0/0 上启用了 LDP，标签签发方式为 DU 方式。
在 Router1 上使用 display mpls ldp session 命令查看 LDP 会话信息。

```
<Router1>display mpls ldp session
 LDP Session(s) in Public Network
 Codes: LAM(Label Advertisement Mode), SsnAge Unit(DDDD:HH:MM)
 A '*' before a session means the session is being deleted.
------------------------------------------------------------------
 PeerID          Status      LAM    SsnRole  SsnAge      KASent/Rcv
------------------------------------------------------------------
```

```
   2.2.2.2:0      Operational  DU   Passive  0000:00:15  61/61
------------------------------------------------------------------
TOTAL: 1 session(s) Found.
```

在 Router2 上使用 **display mpls ldp session** 命令查看 LDP 会话信息。

```
<Router2>display mpls ldp session
------------------------------------------------------------------
PeerID          Status        LAM  SsnRole  SsnAge      KASent/Rcv
------------------------------------------------------------------
1.1.1.1:0       Operational   DU   Active   0000:00:18  76/76
3.3.3.3:0       Operational   DU   Passive  0000:00:13  56/56
------------------------------------------------------------------
TOTAL: 2 session(s) Found.
```

在 Router3 上使用 **display mpls ldp session** 命令查看 LDP 会话信息。

```
<Router3>display mpls ldp session
 ……
------------------------------------------------------------------
PeerID          Status        LAM  SsnRole  SsnAge      KASent/Rcv
------------------------------------------------------------------
2.2.2.2:0       Operational   DU   Active   0000:00:17  72/72
------------------------------------------------------------------
TOTAL: 1 session(s) Found.
```

结果显示，Router1 和 Router2、Router2 和 Router3 之间的 LDP 会话状态为 Operational，表示会话成功建立。

在 Router1 上使用命令 **display mpls lsp** 查看 LSP 信息。

```
<Router1>display mpls lsp
------------------------------------------------------------------
         LSP Information: LDP LSP
------------------------------------------------------------------
FEC             In/Out Label  In/Out IF         Vrf Name
2.2.2.2/32      NULL/3        -/GE0/0/0
2.2.2.2/32      1024/3        -/GE0/0/0
1.1.1.1/32      3/NULL        -/-
3.3.3.3/32      NULL/1025     -/GE0/0/0
3.3.3.3/32      1025/1025     -/GE0/0/0
```

第 11 章　VPN 配置

在 Router2 上使用命令 **display mpls lsp** 查看 LSP 信息。

```
<Router2>display mpls lsp
------------------------------------------------------------------
            LSP Information: LDP LSP
------------------------------------------------------------------
FEC              In/Out Label   In/Out IF           Vrf Name
2.2.2.2/32       3/NULL         -/-
1.1.1.1/32       NULL/3         -/GE0/0/0
1.1.1.1/32       1024/3         -/GE0/0/0
3.3.3.3/32       NULL/3         -/GE0/0/1
3.3.3.3/32       1025/3         -/GE0/0/1
```

在 Router3 上使用命令 **display mpls lsp** 查看 LSP 信息。

```
<Router3>display mpls lsp
------------------------------------------------------------------
            LSP Information: LDP LSP
------------------------------------------------------------------
FEC              In/Out Label   In/Out IF           Vrf Name
1.1.1.1/32       NULL/1024      -/GE0/0/0
1.1.1.1/32       1024/1024      -/GE0/0/0
2.2.2.2/32       NULL/3         -/GE0/0/0
2.2.2.2/32       1025/3         -/GE0/0/0
3.3.3.3/32       3/NULL         -/-
```

结果显示，LDP 为 Router1 到 Router3 以及 Router3 到 Router1 都动态建立了 LSP。

在 Router1 上使用 **tracert lsp ip** 命令查看到主机 3.3.3.3/32 的 MPLS 报文所经过的路径。

```
<Router1>tracert lsp ip 3.3.3.3 32
  LSP Trace Route FEC: IPV4 PREFIX 3.3.3.3/32 , press CTRL_C to break.
 TTL    Replier         Time     Type        Downstream
  0                              Ingress     211.85.1.2/[1025]
  1      211.85.1.2     20 ms    Transit     211.85.2.2/[3]
  2      3.3.3.3        30 ms    Egress
```

在 Router3 上使用 **tracert lsp ip** 命令查看到主机 1.1.1.1/32 的 MPLS 报文所经过的路径。

```
<Router3>tracert lsp ip 1.1.1.1 32
  LSP Trace Route FEC: IPV4 PREFIX 1.1.1.1/32 , press CTRL_C to break.
 TTL    Replier         Time     Type        Downstream
  0                              Ingress     211.85.2.1/[1024]
```

计算机网络实践教程

```
  1    211.85.2.1     10 ms    Transit    211.85.1.1/[3]
  2    1.1.1.1        20 ms    Egress
```

结果显示，动态 LSP 配置和静态 LSP 配置的结果一样（仅标签号不同）。

在 Router1 上使用 **ping lsp ip** 命令测试与 Router3 的连通性。

```
<Router1>ping lsp ip 3.3.3.3 32
  LSP ping FEC: IPV4 PREFIX 3.3.3.3/32/ : 100  data bytes, press CTRL_C to break
    Reply from 3.3.3.3: bytes=100 Sequence=1 time=30 ms
    Reply from 3.3.3.3: bytes=100 Sequence=2 time=20 ms
    Reply from 3.3.3.3: bytes=100 Sequence=3 time=40 ms
    Reply from 3.3.3.3: bytes=100 Sequence=4 time=30 ms
    Reply from 3.3.3.3: bytes=100 Sequence=5 time=20 ms
  --- FEC: IPV4 PREFIX 3.3.3.3/32 ping statistics ---
    5 packet(s) transmitted
    5 packet(s) received
    0.00% packet loss
    round-trip min/avg/max = 20/28/40 ms
```

在 Router3 上使用 **ping lsp ip** 命令测试与 Router1 的连通性。

```
<Router3>ping lsp ip 1.1.1.1 32
  LSP ping FEC: IPV4 PREFIX 1.1.1.1/32/ : 100  data bytes, press CTRL_C to break
    Reply from 1.1.1.1: bytes=100 Sequence=1 time=20 ms
    Reply from 1.1.1.1: bytes=100 Sequence=2 time=30 ms
    Reply from 1.1.1.1: bytes=100 Sequence=3 time=20 ms
    Reply from 1.1.1.1: bytes=100 Sequence=4 time=40 ms
    Reply from 1.1.1.1: bytes=100 Sequence=5 time=30 ms
  --- FEC: IPV4 PREFIX 1.1.1.1/32 ping statistics ---
    5 packet(s) transmitted
    5 packet(s) received
    0.00% packet loss
    round-trip min/avg/max = 20/28/40 ms
```

结果显示，Router1 与 Router3 之间可以通过 MPLS 的 LSP 进行报文转发。

第 11 章 VPN 配置

11.5 BGP/MPLS VPN 基本配置

一、原理概述

BGP/MPLS VPN 也简称 MPLS L3 VPN，是 ISP 的 VPN 解决方案中的一种基于 PE 的 L3 VPN 技术，它使用 BGP 在服务提供商骨干网上发布 VPN 路由，使用 MPLS 在服务提供商骨干网上转发 VPN 报文。

MPLS L3 VPN 组网方式灵活、可扩展性好，并能够方便地支持 MPLS QoS 和 MPLS TE，因此得到越来越多的应用。

MPLS L3 VPN 模型由三部分组成：CE（customer edge）、PE（provider edge）和 P（provider）。

(1) CE 设备：用户网络边缘设备，有接口直接与服务提供商（service provider，SP）相连。CE 可以是路由器、交换机或主机。CE "感知" 不到 VPN 的存在，也不需要必须支持 MPLS。

(2) PE 路由器：服务提供商边缘路由器，是服务提供商网络的边缘设备，与用户的 CE 直接相连。在 MPLS 网络中，对 VPN 的所有处理都发生在 PE 上。

(3) P 路由器：服务提供商网络中的骨干路由器，不与 CE 直接相连。P 设备只需要具备基本 MPLS 转发能力。

二、实验目的

1. 掌握 BGP 和 MPLS 的工作原理
2. 掌握 BGP/MPLS VPN 的工作过程

三、实验内容

本实验模拟企业网络环境，分处两地的公司 A 的两个子网（分别用 Router1 和 Router2 表示）通过 VPN 骨干网互联，分处两地的公司 B 的两个子网（分别由 Router3 和 Router4 表示）通过 VPN 骨干网互联。VPN 骨干网通过 OSPF 互联；公司 A 的 CE（Router1 和 Router2）设备通过 BGP 与 PE（Router5 和 Router7）设备互联；公司 B 的两个子网，Router3 和 Router5 采用静态路由的方式传递路由，Router4 与 Router7 采用 OSPF 传递路由；Router6 通过 OSPF 将私网路由传递给 PE 设备。

四、实验拓扑

BGP/MPLS VPN 配置拓扑如图 11.11 所示。

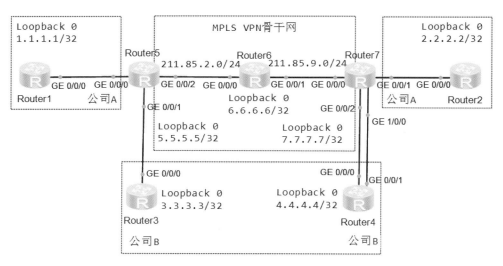

图 11.11　BGP/MPLS VPN 配置拓扑

五、实验地址分配

实验地址分配如表 11.5 所示。

表 11.5　实验地址分配

设备	接口	IP 地址	子网掩码	默认网关
Router1 （AR2220）	GE 0/0/0	192.168.10.1	255.255.255.0	N/A
	Loopback 0	1.1.1.1	255.255.255.255	N/A
Router2 （AR2220）	GE 0/0/0	192.168.20.2	255.255.255.0	N/A
	Loopback 0	2.2.2.2	255.255.255.255	N/A
Router3 （AR2220）	GE 0/0/0	192.168.30.3	255.255.255.0	N/A
	Loopback 0	3.3.3.3	255.255.255.255	N/A
Router4 （AR2220）	GE 0/0/0	192.168.40.4	255.255.255.0	N/A
	GE 0/0/1	211.85.11.4	255.255.255.0	N/A
	Loopback 0	4.4.4.4	255.255.255.255	N/A
Router5 （AR2220）	GE 0/0/0	192.168.10.5	255.255.255.0	N/A
	GE 0/0/1	192.168.30.5	255.255.255.0	N/A
	GE 0/0/2	211.85.2.5	255.255.255.0	N/A
	Loopback 0	5.5.5.5	255.255.255.255	N/A
Router6 （AR2220）	GE 0/0/0	211.85.2.6	255.255.255.0	N/A
	GE 0/0/1	211.85.9.6	255.255.255.0	N/A
	Loopback 0	6.6.6.6	255.255.255.255	N/A

续表

设备	接口	IP 地址	子网掩码	默认网关
Router7 （AR2220）	GE 0/0/0	211.85.9.7	255.255.255.0	N/A
	GE 0/0/1	192.168.20.7	255.255.255.0	N/A
	GE 0/0/2	192.168.40.7	255.255.255.0	N/A
	GE 1/0/0	211.85.11.7	255.255.255.0	N/A
	Loopback 0	7.7.7.7	255.255.255.255	N/A

六、实验步骤

1. 基本配置

根据实验地址分配表进行各设备的相应配置。

1) Router1 的基本配置

```
<Huawei>system-view
[Huawei]sysname Router1
[Router1]interface GigabitEthernet 0/0/0
[Router1-GigabitEthernet0/0/0]ip address 192.168.10.1 24
[Router1-GigabitEthernet0/0/0]interface loopback 0
[Router1-LoopBack0]ip address 1.1.1.1 32
```

2) Router2 的基本配置

```
<Huawei>system-view
[Huawei]sysname Router2
[Router2]interface GigabitEthernet 0/0/0
[Router2-GigabitEthernet0/0/0]ip address 192.168.20.2 24
[Router2-GigabitEthernet0/0/0]interface loopback 0
[Router2-LoopBack0]ip address 2.2.2.2 32
```

3) Router3 的基本配置

```
<Huawei>system-view
[Huawei]sysname Router3
[Router3]interface GigabitEthernet 0/0/0
[Router3-GigabitEthernet0/0/0]ip address 192.168.30.3 24
[Router3-GigabitEthernet0/0/0]interface loopback 0
[Router3-LoopBack0]ip address 3.3.3.3 32
```

4) Router4 的基本配置

```
<Huawei>system-view
[Huawei]sysname Router4
[Router4]interface GigabitEthernet 0/0/0
[Router4-GigabitEthernet0/0/0]ip address 192.168.40.4 24
[Router4-GigabitEthernet0/0/0]interface GigabitEthernet 0/0/1
[Router4-GigabitEthernet0/0/1]ip address 211.85.11.4 24
[Router4-GigabitEthernet0/0/1]interface looPback 0
[Router4-LoopBack0]ip address 4.4.4.4 32
```

5) Router5 的基本配置

```
<Huawei>system-view
[Huawei]sysname Router5
[Router5]interface GigabitEthernet 0/0/0
[Router5-GigabitEthernet0/0/0]ip address 192.168.10.5 24
[Router5-GigabitEthernet0/0/0]interface GigabitEthernet 0/0/1
[Router5-GigabitEthernet0/0/1]ip address 192.168.30.5 24
[Router5-GigabitEthernet0/0/1]interface GigabitEthernet 0/0/2
[Router5-GigabitEthernet0/0/2]ip address 211.85.2.5 24
[Router5-GigabitEthernet0/0/2]interface loopback 0
[Router5-LoopBack0]ip address 5.5.5.5 32
```

6) Router6 的基本配置

```
<Huawei>system-view
[Huawei]sysname Router6
[Router6]interface GigabitEthernet 0/0/0
[Router6-GigabitEthernet0/0/0]ip address 211.85.2.6 24
[Router6-GigabitEthernet0/0/0]interface GigabitEthernet 0/0/1
[Router6-GigabitEthernet0/0/1]ip address 211.85.9.6 24
[Router6-GigabitEthernet0/0/1]interface loopback 0
[Router6-LoopBack0]ip address 6.6.6.6 32
```

7) Router7 的基本配置

```
<Huawei>system-view
[Huawei]sysname Router7
[Router7]interface GigabitEthernet 0/0/0
[Router7-GigabitEthernet0/0/0]ip address 211.85.9.7 24
[Router7-GigabitEthernet0/0/0]interface GigabitEthernet 0/0/1
```

```
[Router7-GigabitEthernet0/0/1]ip address 192.168.20.7 24
[Router7-GigabitEthernet0/0/1]interface GigabitEthernet 0/0/2
[Router7-GigabitEthernet0/0/2]ip address 192.168.40.7 24
[Router7-GigabitEthernet0/0/2]interface GigabitEthernet 1/0/0
[Router7-GigabitEthernet1/0/0]ip address 211.85.11.7 24
[Router7-GigabitEthernet1/0/0]interface loopback 0
[Router7-LoopBack0]ip address 7.7.7.7 32
```

8) 测试相连路由器之间的连通性

在 Router5 上使用 ping 命令测试与 Router6 的连通性。使用同样的方法测试其他相连路由器之间的连通性。

```
<Router5>ping -c 1 211.85.2.6
 ping 211.85.2.6: 56  data bytes, press CTRL_C to break
   Reply from 211.85.2.6: bytes=56 Sequence=1 ttl=255 time=110 ms
 --- 211.85.2.6 ping statistics ---
   1 packet(s) transmitted
   1 packet(s) received
   0.00% packet loss
   round-trip min/avg/max = 20/42/110 ms
```

结果显示，Router5 与 Router6 正常通信。

2. 在 MPLS VPN 骨干网上配置 OSPF 网络

在骨干网上使用内部网关协议 OSPF 配置路由，各路由器均属于区域 0，使用 Loopback 0 接口 IP 地址作为 Router-ID。

(1) 在 Router5 上配置 OSPF 网络，Router-ID 为 5.5.5.5。

```
[Router5]ospf router-id 5.5.5.5
[Router5-ospf-1]area 0
[Router5-ospf-1-area-0.0.0.0]network 211.85.2.0 0.0.0.255
[Router5-ospf-1-area-0.0.0.0]network 5.5.5.5 0.0.0.0
```

(2) 在 Router6 上配置 OSPF 网络，Router-ID 为 6.6.6.6。

```
[Router6]ospf router-id 6.6.6.6
[Router6-ospf-1]area 0
[Router6-ospf-1-area-0.0.0.0]network 211.85.2.0 0.0.0.255
[Router6-ospf-1-area-0.0.0.0]network 211.85.9.0 0.0.0.255
[Router6-ospf-1-area-0.0.0.0]network 6.6.6.6 0.0.0.0
```

(3) 在 Router7 上配置 OSPF 网络，Router-ID 为 7.7.7.7。

```
[Router7]ospf router-id 7.7.7.7
[Router7-ospf-1]area 0
[Router7-ospf-1-area-0.0.0.0]network 211.85.9.0 0.0.0.255
[Router7-ospf-1-area-0.0.0.0]network 7.7.7.7 0.0.0.0
```

(4) 查看 OSPF 邻居信息。

在 Router6 上使用 **display ospf peer brief** 命令查看 OSPF 邻居信息。

```
[Router6]display ospf peer brief
 OSPF Process 1 with Router ID 6.6.6.6
      Peer Statistic Information
 ----------------------------------------------------------
 Area Id     Interface               Neighbor id      State
 0.0.0.0     GigabitEthernet0/0/0    5.5.5.5          Full
 0.0.0.0     GigabitEthernet0/0/1    7.7.7.7          Full
 ----------------------------------------------------------
```

结果显示，Router6 与 Router5、Router7 的 OSPF 邻居状态为 Full，表明邻居关系建立成功。

3. 在骨干网上配置 MPLS 与 LDP

在骨干网内配置 MPLS 和 LDP，各路由器使用 Loopback 0 接口地址作为 LSR-ID。

(1) 在 Router5 上使用 **mpls ldp** 命令配置 MPLS。

```
[Router5]mpls lsr-id 5.5.5.5
[Router5]mpls
[Router5-mpls]mpls ldp
[Router5-mpls-ldp]interface GigabitEthernet 0/0/2
[Router5-GigabitEthernet0/0/2]mpls
[Router5-GigabitEthernet0/0/2]mpls ldp
```

(2) 在 Router7 上使用 **mpls ldp** 命令配置 MPLS。

```
[Router7]mpls lsr-id 7.7.7.7
[Router7]mpls
[Router7-mpls]mpls ldp
[Router7-mpls-ldp]interface GigabitEthernet 0/0/0
[Router7-GigabitEthernet0/0/0]mpls
[Router7-GigabitEthernet0/0/0]mpls ldp
```

第 11 章　VPN 配置

(3) 在 Router6 上使用 mpls ldp 命令配置 MPLS。

```
[Router6]mpls lsr-id 6.6.6.6
[Router6]mpls
[Router6-mpls]mpls ldp
[Router6-mpls-ldp]interface GigabitEthernet 0/0/0
[Router6-GigabitEthernet0/0/0]mpls
[Router6-GigabitEthernet0/0/0]mpls ldp
[Router6-GigabitEthernet0/0/0]interface GigabitEthernet 0/0/1
[Router6-GigabitEthernet0/0/1]mpls
[Router6-GigabitEthernet0/0/1]mpls ldp
```

(4) 查看 LDP 会话信息。

配置完成后，在 Router6 上使用 display mpls ldp session 命令查看 LDP 会话建立情况。

```
[Router6]display mpls ldp session
 LDP Session(s) in Public Network
 Codes: LAM(Label Advertisement Mode), SsnAge Unit(DDDD:HH:MM)
 A '*' before a session means the session is being deleted.
------------------------------------------------------------------
 PeerID          Status      LAM  SsnRole   SsnAge      KASent/Rcv
------------------------------------------------------------------
 5.5.5.5:0       Operational DU   Active    0000:00:05  23/23
 7.7.7.7:0       Operational DU   Passive   0000:00:00  4/4
------------------------------------------------------------------
 TOTAL: 2 session(s) Found.
```

结果显示，LDP 会话状态为 Operational，会话建立成功。

(5) 查看路由器 LSP 信息。

在 Router5 上使用 display mpls lsp 命令查看 LSP 信息。

```
[Router5]display mpls lsp
----------------------------------------------------
            LSP Information: LDP LSP
----------------------------------------------------
FEC              In/Out Label    In/Out IF      Vrf Name
6.6.6.6/32       NULL/3          -/GE0/0/2
6.6.6.6/32       1024/3          -/GE0/0/2
5.5.5.5/32       3/NULL          -/-
7.7.7.7/32       NULL/1025       -/GE0/0/2
7.7.7.7/32       1025/1025       -/GE0/0/2
```

结果显示，MPLS 网络已经为 Router5、Router6 和 Router7 的 Loopback 0 接口路由建立了相应的 LSP。

4. 配置 PE 设备间的 MP-BGP

(1) 在 Router5 上建立 Router5 与 Router7 的 IBGP 邻居关系。

在 Router5 上使用 bgp 和 peer 命令建立 IBGP 邻居关系，然后使用 **ipv4-family vpnv4** 命令进入 VPNv4 视图，在 VPNv4 视图下启用与 Router7 交换 VPNv4 路由的功能，最后允许与 Router7 交换路由信息时携带 BGP 团体属性。

```
[Router5]bgp 100
[Router5]router-id 5.5.5.5
[Router5-bgp]peer 7.7.7.7 as-number 100
[Router5-bgp]peer 7.7.7.7 connect-interface loopback 0
[Router5-bgp]peer 7.7.7.7 next-hop-local
[Router5-bgp]ipv4-family vpnv4
[Router5-bgp-af-vpnv4]peer 7.7.7.7 enable
[Router5-bgp-af-vpnv4]peer 7.7.7.7 advertise-community
```

(2) 在 Router7 上建立 Router7 与 Router5 的 IBGP 邻居关系。

```
[Router7]bgp 100
[Router7]router-id 7.7.7.7
[Router7-bgp]peer 5.5.5.5 as-number 100
[Router7-bgp]peer 5.5.5.5 connect-interface loopback 0
[Router7-bgp]peer 5.5.5.5 next-hop-local
[Router7-bgp]ipv4-family vpnv4
[Router7-bgp-af-vpnv4]peer 5.5.5.5 enable
[Router7-bgp-af-vpnv4]peer 5.5.5.5 advertise-community
```

(3) 查看 BGP 邻居关系。

在 Router5 上使用 **display bgp peer** 命令查看 BGP 邻居关系。

```
[Router5]display bgp peer
 BGP local router ID : 192.168.10.5
 Local AS number : 100
 Total number of peers:1    Peers in established state : 1
 Peer      V AS  MsgRcvd MsgSent  OutQ  Up/Down   State      PrefRcv
 7.7.7.7   4 100 4       5        0     00:02:26  Established 0
```

结果显示，Router5 与 Router7 之间的邻居状态为 Established，表明 BGP 邻居关系建立成功。

5. 在 PE 上创建 VPN 实例并与接口进行绑定

首先分别在 PE、ASBR 设备上建立 VPN 实例，在 VPN 实例中启用 IPv4 地址簇，然后进入 IPv4 地址簇视图中配置 RD、RT 等；最后配置 import、Export 方向的 VPN-Target 团体属性。

(1) 在 Router5 上创建 VPN 实例并与接口进行绑定。

在 Router5 上使用 **ip vpn-instance** 命令为公司 A 创建名为 vpn-a 的 VPN 实例，进入实例视图；在实例视图下使用 **ipv4-family** 命令启用 VPN 实例的 IPv4 地址簇，进入 IPv4 地址簇视图，在 IPv4 地址簇视图下使用 **route-distinguisher 300:1** 命令配置 RD 为 300:1。最后使用 **vpn-target 100:1 both** 命令配置 Import 和 Export 方向的 VPN-Target 团体属性，值均为 100:1。

```
[Router5]ip vpn-instance vpn-a
[Router5-vpn-instance-vpn-a]ipv4-family
[Router5-vpn-instance-vpn-a-af-ipv4]route-distinguisher 300:1
[Router5-vpn-instance-vpn-a-af-ipv4]vpn-target 100:1 both
```

在 Router5 连接公司 A 的 GE 0/0/0 接口的视图下，使用 **ip binding vpn-instance vpn-a** 命令将 GE 0/0/0 接口与 VPN 实例 vpn-a 进行绑定。

```
[Router5]interface GigabitEthernet 0/0/0
[Router5-GigabitEthernet0/0/0]ip binding vpn-instance vpn-a
Info: All IPv4 related configurations on this interface are removed!
Info: All IPv6 related configurations on this interface are removed!
[Router5-GigabitEthernet0/0/0]ip address 192.168.10.5 24
```

注意，绑定后的接口的 IP 地址等信息将被删除，需要重新配置。

同样，在 Router5 上为公司 B 创建名为 vpn-b 的 VPN 实例，RD 为 300:2，VPN-Targe 为 100:2，绑定接口为 GE 0/0/1。

```
[Router5]ip vpn-instance vpn-b
[Router5-vpn-instance-vpn-b]ipv4-family
[Router5-vpn-instance-vpn-b-af-ipv4]route-distinguisher 300:2
[Router5-vpn-instance-vpn-b-af-ipv4]vpn-target 100:2 both
[Router5-vpn-instance-vpn-b-af-ipv4]interface GigabitEthernet 0/0/1
[Router5-GigabitEthernet0/0/1]ip binding vpn-instance vpn-b
[Router5-GigabitEthernet0/0/1]ip address 192.168.30.5 24
```

(2) 在 Router7 上创建 VPN 实例并与接口进行绑定。

在 Router7 上进行同样的操作。

```
[Router7]ip vpn-instance vpn-a
[Router7-vpn-instance-vpn-a]ipv4-family
[Router7-vpn-instance-vpn-a-af-ipv4]route-distinguisher 300:1
[Router7-vpn-instance-vpn-a-af-ipv4]vpn-targe 100:1 both
[Router7]ip vpn-instance vpn-b
[Router7-vpn-instance-vpn-b]ipv4-family
[Router7-vpn-instance-vpn-b-af-ipv4]route-distinguisher 300:2
[Router7-vpn-instance-vpn-b-af-ipv4]vpn-targe 100:2 both
[Router7-vpn-instance-vpn-a-af-ipv4]interface GigabitEthernet 0/0/0
[Router7-GigabitEthernet0/0/0]ip binding vpn-instance vpn-a
[Router7-GigabitEthernet0/0/0]ip address 211.85.9.7 24
[Router7-GigabitEthernet0/0/0]interface GigabitEthernet 0/0/2
[Router7-GigabitEthernet0/0/2]ip binding vpn-instance vpn-b
[Router7-GigabitEthernet0/0/2]ip address 192.168.40.7 24
```

6. 为公司 A 配置基于 BGP 的 PE 与 CE 连通性

在公司 A 的 CE 设备 Router1 上进行 BGP 配置,建立与 PE 设备 Router1 的 EBGP 邻居关系。

```
[Router1]bgp 10
[Router1-bgp]peer 192.168.10.5 as-number 100
[Router1-bgp]network 1.1.1.1 32
```

在 PE 设备 Router5 的 BGP 视图下使用 ipv4-family vpn-instance vpn-a 命令进入 VPN 实例 vpn-a 的视图,然后与 Router1 建立 BGP 邻居关系。

```
[Router5]bgp 100
[Router5-bgp]ipv4-family vpn-instance vpn-a
[Router5-bgp-vpn-a]peer 192.168.10.1 as-number 10
```

配置完成后,在 Router1 上使用 display bgp peer 命令查看 BGP 邻居状态。

```
[Router1]display bgp peer
 BGP local router ID : 192.168.10.1
 Local AS number : 10
 Total number of peers : 1      Peers in established state : 1
 Peer            V    AS   MsgRcvd  MsgSent  OutQ  Up/Down    State PrefRcv
 192.168.10.5    4    100    12      14       0    00:10:59   Established
```

结果显示,Router1 与 192.168.10.5 的 BGP 邻居关系状态为 Established,邻居关系建立成功。

第 11 章 VPN 配置

在 Router5 上使用 display bgp peer 命令查看 BGP 邻居关系。

```
[Router5]display bgp peer
BGP local router ID : 192.168.10.5
Local AS number : 100
Total number of peers : 2      Peers in established state : 0
  Peer            V  AS  MsgRcvd  MsgSent  OutQ  Up/Down   State       PrefRcv
  7.7.7.7         4  100    2       4       0   00:00:10  Established     0
  192.168.10.1    4  10     0       0       0   00:20:55  Idle            0
```

结果显示，在 Router5 上并不能查到 192.168.10.1 的邻居信息，正确的方法是使用 display bgp vpnv4 vpn-instance vpn-a peer 命令。

```
[Router5]display bgp vpnv4 vpn-instance vpn-a peer
BGP local router ID : 192.168.10.5
Local AS number : 100
VPN-Instance vpn-a, Router ID 192.168.10.5:
Total number of peers : 1      Peers in established state : 1
  Peer            V  AS  MsgRcvd  MsgSent  OutQ  Up/Down   State       PrefRcv
  192.168.10.1    4  10   6773     6772     0    0112h50m  Established     1
```

结果显示，Router5 与 192.168.10.1 的邻居关系建立成功。
在 Router2 与 Router7 上完成相同的配置。

```
[Router2]bgp 20
[Router2-bgp]peer 192.168.20.7 as-number 100
[Router2-bgp]network 2.2.2.2 32

[Router7]bgp 100
[Router7-bgp]ipv4-family vpn-instance vpn-a
[Router7-bgp-vpn-a]peer 192.168.20.2 as-number 20
```

在 Router5 上使用 display bgp vpnv4 vpn-instance vpn-a routing-table 命令查看 VPN 实例 vpn-a 的 BGP 路由表。

```
[Router5]display bgp vpnv4 vpn-instance vpn-a routing-table
BGP Local router ID is 211.85.2.5
Status codes: * - valid, > - best, d - damped,
              h - history, i - internal, s - suppressed, S - Stale
              Origin : i - IGP, e - EGP, ? - incomplete
VPN-Instance vpn-a, Router ID 211.85.2.5:
```

计算机网络实践教程

```
Total Number of Routes: 2
     Network        NextHop          MED      LocPrf   PrefVal   Path/Ogn
 *>  1.1.1.1/32     192.168.10.1     0                 0         10i
 *>i 2.2.2.2/32     7.7.7.7          0        100      0         20i
```

结果显示，VPN 实例 vpn-a 仅仅拥有 1.1.1.1/32 和 2.2.2.2/32 的路由。

在 Router5 上使用 **display mpls lsp** 命令查看 LSP 信息。

```
[Router5]display mpls lsp
------------------------------------------------------------
              LSP Information: BGP  LSP
------------------------------------------------------------
FEC              In/Out Label    In/Out IF         Vrf Name
1.1.1.1/32       1026/NULL       -/-               vpn-a
------------------------------------------------------------
              LSP Information: LDP LSP
------------------------------------------------------------
FEC              In/Out Label    In/Out IF         Vrf Name
5.5.5.5/32       3/NULL          -/-
6.6.6.6/32       NULL/3          -/GE0/0/2
6.6.6.6/32       1024/3          -/GE0/0/2
7.7.7.7/32       NULL/1025       -/GE0/0/2
7.7.7.7/32       1025/1025       -/GE0/0/2
```

结果显示了 BGP LSP 的信息，FEC 为 1.1.1.1/32，In 标签为 1026，Out 标签为 NULL，VRF Name 为 vpn-a。In 标签 1026 应该是由 MP-BGP 协议分配的内层标签，仅用于区分路由信息所属的 VRF。

在 Router7 上使用 **display mpls lsp** 命令查看 LSP 信息。

```
[Router7]display mpls lsp
------------------------------------------------------------
              LSP Information: BGP  LSP
------------------------------------------------------------
FEC              In/Out Label    In/Out   IF        Vrf Name
2.2.2.2/32       1026/NULL       -/-                vpn-a
------------------------------------------------------------
              LSP Information: LDP LSP
------------------------------------------------------------
FEC              In/Out Label    In/Out   IF        Vrf Name
5.5.5.5/32       NULL/1024       -/GE0/0/0
5.5.5.5/32       1024/1024       -/GE0/0/0
6.6.6.6/32       NULL/3          -/GE0/0/0
6.6.6.6/32       1025/3          -/GE0/0/0
```

第 11 章 VPN 配置

```
7.7.7.7/32      3/NULL          -/-
```

结果显示，对于 FEC 2.2.2.2/32，MP-BGP 分配的标签为 1026。

在 Router1 上以 1.1.1.1 为源，使用 ping 命令测试与 2.2.2.2 的连通性。

```
<Router1>ping -a 1.1.1.1 2.2.2.2
 ping 2.2.2.2: 56  data bytes, press CTRL_C to break
   Reply from 2.2.2.2: bytes=56 Sequence=1 ttl=252 time=90 ms
   Reply from 2.2.2.2: bytes=56 Sequence=2 ttl=252 time=30 ms
   Reply from 2.2.2.2: bytes=56 Sequence=3 ttl=252 time=30 ms
   Reply from 2.2.2.2: bytes=56 Sequence=4 ttl=252 time=40 ms
   Reply from 2.2.2.2: bytes=56 Sequence=5 ttl=252 time=30 ms
 --- 2.2.2.2 ping statistics ---
   5 packet(s) transmitted
   5 packet(s) received
   0.00% packet loss
   round-trip min/avg/max = 30/44/90 ms
```

结果显示，Router1 和 Router2 能够正常通信。实现了公司 A 的 VPN 网络的通信。

7. 为公司 B 配置基于静态路由及 OSPF 协议的 PE 与 CE 连通

（1）公司 B 的 CE 设备 Router3 使用静态路由实现 PE 与 CE 连通。

在 Router3 上配置默认路由。

```
[Router3]ip route-static 0.0.0.0 0 192.168.30.5
```

在 Router5 上为 VPN 实例 vpn-b 配置静态路由。

```
[Router5]ip route-static vpn-instance vpn-b 3.3.3.3 32 192.168.30.3
```

在 Router5 上使用 ipv4-family vpn-instance vpn-b 命令进入 VPN 实例视图，然后将 VPN 实例 vpn-b 的静态路由引入 BGP。

```
[Router5]bgp 100
[Router5-bgp]ipv4-family vpn-instance vpn-b
[Router5-bgp-vpn-b]import-route static
```

（2）公司 B 的 CE 设备 Router4 使用 OSPF 协议实现 PE 与 CE 连通。

在 Router4 上进行 OSPF 配置。

```
[Router4]ospf 2 router-id 4.4.4.4
[Router4-ospf-2]area 0
[Router4-ospf-2-area-0.0.0.0]network 192.168.40.0 0.0.0.255
[Router4-ospf-2-area-0.0.0.0]network 4.4.4.4 0.0.0.0
```

在 Router7 上为 VPN 实例 vpn-b 配置 OSPF。

```
[Router7]ospf 2 vpn-instance vpn-b
[Router7-ospf-2]area 0
[Router7-ospf-2-area-0.0.0.0]network 192.168.40.0 0.0.0.255
```

在 Router7 上使用 **display ospf peer brief** 命令查看 OSPF 邻居状态。

```
[Router7]display ospf peer brief
 OSPF Process 1 with Router ID 7.7.7.7
     Peer Statistic Information
 ----------------------------------------------------------------
 Area Id     Interface              Neighbor id     State
 0.0.0.0     GigabitEthernet0/0/0   6.6.6.6         Full
 ----------------------------------------------------------------
 OSPF Process 2 with Router ID 192.168.40.7
     Peer Statistic Information
 ----------------------------------------------------------------
 Area Id     Interface              Neighbor id     State
 0.0.0.0     GigabitEthernet0/0/2   4.4.4.4         Full
 ----------------------------------------------------------------
```

结果显示，Router7 与 Router4 的邻居状态为 Full，邻居关系建立成功。

在 Router7 的 OSPF 视图下使用 import-route bgp 命令将 VPN 实例 vpn-b 的 BGP 路由引入 OSPF。

```
[Router7]ospf 2
[Router7-ospf-2]import-route bgp
```

在 Router7 的 BGP 视图下使用 ipv4-family vpn-instance vpn-b 命令引入 VPN 实例 vpn-b 的视图，将 VPN 实例 vpn-b 的 OSPF 路由引入 BGP。

```
[Router7]bgp 100
[Router7-bgp]ipv4-family vpn-instance vpn-b
```

第 11 章 VPN 配置

```
[Router7-bgp-vpn-b]import-route ospf 2
```

(3) 测试。

在 Router7 上使用 **display bgp vpnv4 vpn-instance vpn-b routing-table** 命令查看 VPN 实例 vpn-b 的 BGP 路由表。

```
[Router7]display bgp vpnv4 vpn-instance vpn-b routing-table
BGP Local router ID is 211.85.9.7
Status codes: * - valid, > - best, d - damped,
              h - history, i - internal, s - suppressed, S - Stale
              Origin : i - IGP, e - EGP, ? - incomplete
VPN-Instance vpn-b, Router ID 211.85.9.7:
Total Number of Routes: 3
     Network            NextHop         MED      LocPrf    PrefVal Path/Ogn
 *>i 3.3.3.3/32         5.5.5.5         0        100       0        ?
 *>  4.4.4.4/32         0.0.0.0         2                  0        ?
 *>  192.168.40.0       0.0.0.0         0                  0        ?
```

结果显示，Router7 的 BGP 路由表中不仅有 3.3.3.3/32 和 4.4.4.4/32 的路由信息，而且还有 192.168.40.0/24 的路由信息。与使用 BGP 协议或静态路由方式来提供 PE-CE 连通性不同，使用 OSPF 协议实现 PE-CE 的连通性时，PE 与 CE 之间的链路在引入路由时也会引入 BGP。

在 Router7 上使用 **display mpls lsp** 命令查看 LSP。

```
[Router7]display mpls lsp
-------------------------------------------------------------
        LSP Information: BGP  LSP
-------------------------------------------------------------
FEC                  In/Out Label    In/Out IF       Vrf Name
2.2.2.2/32           1026/NULL       -/-             vpn-a
4.4.4.4/32           1027/NULL       -/-             vpn-b
192.168.40.0/24      1028/NULL       -/-             vpn-b
        LSP Information: LDP  LSP
-------------------------------------------------------------
FEC                  In/Out Label    In/Out IF       Vrf Name
5.5.5.5/32           NULL/1024       -/GE0/0/0
5.5.5.5/32           1024/1024       -/GE0/0/0
6.6.6.6/32           NULL/3          -/GE0/0/0
6.6.6.6/32           1025/3          -/GE0/0/0
7.7.7.7/32           3/NULL          -/-
```

结果显示，MP-BGP 协议为 4.4.4.4/32 和 192.168.40.0/24 分配了标签。

在 Router4 上以 4.4.4.4 为源，使用 **ping** 命令测试与 1.1.1.1/32、2.2.2.2/32、3.3.3.3/32 之间

的连通性。

测试与公司 A 的 1.1.1.1/32 的连通性。

```
<Router4>ping -a 4.4.4.4 1.1.1.1
  ping 1.1.1.1: 56  data bytes, press CTRL_C to break
    Request time out
    Request time out
    Request time out
    Request time out
    Request time out
  --- 1.1.1.1 ping statistics ---
    5 packet(s) transmitted
    0 packet(s) received
    100.00% packet loss
```

测试与公司 A 的 2.2.2.2/32 的连通性。

```
<Router4>ping -a 4.4.4.4 2.2.2.2
  ping 2.2.2.2: 56  data bytes, press CTRL_C to break
    Request time out
    Request time out
    Request time out
    Request time out
    Request time out
  --- 2.2.2.2 ping statistics ---
    5 packet(s) transmitted
    0 packet(s) received
    100.00% packet loss
```

测试与公司 B 的 3.3.3.3/32 的连通性。

```
<Router4>ping -a 4.4.4.4 3.3.3.3
  ping 3.3.3.3: 56  data bytes, press CTRL_C to break
    Reply from 3.3.3.3: bytes=56 Sequence=1 ttl=252 time=30 ms
    Reply from 3.3.3.3: bytes=56 Sequence=2 ttl=252 time=30 ms
    Reply from 3.3.3.3: bytes=56 Sequence=3 ttl=252 time=30 ms
    Reply from 3.3.3.3: bytes=56 Sequence=4 ttl=252 time=30 ms
    Reply from 3.3.3.3: bytes=56 Sequence=5 ttl=252 time=30 ms
  --- 3.3.3.3 ping statistics ---
    5 packet(s) transmitted
    5 packet(s) received
    0.00% packet loss
    round-trip min/avg/max = 30/30/30 ms
```

结果显示，Router4 仅能够与同属公司 B 的 3.3.3.3/32 进行通信，不能与属于公司 A 的 1.1.1.1/32 和 2.2.2.2/32 进行通信。

当 CE-PE 之间运行 EBGP 时，无需在 PE 上对客户路由和 MP-BGP 协议之间进行引入配置，客户的 VPNv4 路由可以直接通过 MPLS/MP-BGP 网络传递给对端 PE。而当 CE-PE 之间运行的是静态路由或 IGP 时，则需要进行互相引入的配置，才能使客户的 VPN4 的路由通过 MPLS/MP-BGP 网络进行传递。

第 12 章 防火墙配置

一、原理概述

防火墙指的是一个由软件与硬件设备组成、在内部网络与外部网络之间、专用网与公共网之间构造的保护隔离屏障,是一种获取安全性方法的形象说法。它是一种计算机硬件和软件的结合,使 Internet 与 Intranet 之间建立起一个安全网关,从而保护内部网络免受非法用户的侵入。防火墙主要由服务访问规则、验证工具、包过滤和应用网关 4 个部分组成,防火墙就是一个位于计算机和它所连接的网络之间的软件或硬件。该计算机流入/流出的所有网络通信和数据包均要经过此防火墙。

防火墙的工作模式主要有 3 种:路由模式、透明模式和混合模式。

(1) 路由模式是指设备接口具有 IP 地址,通过 3 层对外连接;
(2) 透明模式是指设备接口没有 IP 地址,通过 2 层对外连接;
(3) 混合模式是指设备接口既有工作在路由模式的接口,又有工作在透明模式的接口。

二、实验目的

1. 理解防火墙的应用场景
2. 理解防火墙的基本原理和工作模式
3. 掌握防火墙的配置过程

三、实验内容

本实验模拟企业网络环境,某公司因业务需要,现要求异地子公司能够通过 Internet 把子公司关键业务数据安全地传送给总公司。要求子公司可以访问总公司的 Web 服务器、FTP 服务器、Telnet 服务器。总公司通过防火墙连接 Internet,子公司通过路由器连接到 Internet。使用防火墙解决这个问题。

四、实验拓扑

防火墙配置拓扑如图 12.1 所示。

第 12 章 防火墙配置

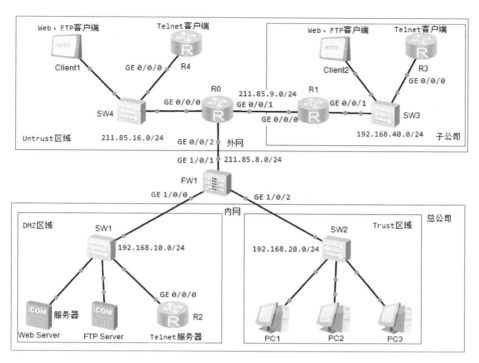

图 12.1 防火墙配置拓扑

五、实验地址分配

实验地址分配如表 12.1 所示。

表 12.1 实验地址分配

设备	接口	IP 地址	子网掩码	默认网关
FW1 (USG6000V)	GE 1/0/0	192.168.10.254		
	GE 1/0/1	211.85.8.1		
	GE 1/0/2	192.168.20.254		
R0（AR2220）	GE 0/0/0	211.85.16.254	255.255.255.0	N/A
	GE 0/0/1	211.85.9.1	255.255.255.0	N/A
	GE 0/0/2	211.85.8.2	255.255.255.0	N/A
R1（AR2220）	GE 0/0/0	211.85.9.2	255.255.255.0	N/A
	GE 0/0/1	192.168.40.254	255.255.255.0	N/A
R2（AR2220）	GE 0/0/0	192.168.10.3	255.255.255.0	N/A
R3（AR2220）	GE 0/0/0	192.168.40.2	255.255.255.0	N/A
R4（AR2220）	GE 0/0/0	211.85.16.2	255.255.255.0	N/A
Web Server	Ethernet 0/0/0	192.168.10.1	255.255.255.0	192.168.10.254

续表

设备	接口	IP 地址	子网掩码	默认网关
FTP Server	Ethernet 0/0/0	192.168.10.2	255.255.255.0	192.168.10.254
Client1	Ethernet 0/0/0	211.85.16.1	255.255.255.0	211.85.16.254
Client2	Ethernet 0/0/0	192.168.40.1	255.255.255.0	192.168.40.254
PC1	Ethernet 0/0/1	192.168.20.1	255.255.255.0	192.168.20.254
PC2	Ethernet 0/0/1	192.168.20.2	255.255.255.0	192.168.20.254
PC3	Ethernet 0/0/1	192.168.20.3	255.255.255.0	192.168.20.254

六、实验步骤

1. PC 的配置

1) PC1 的配置

根据实验地址分配，PC1 的配置如图 12.2 所示。

图 12.2 PC1 的配置

2) PC2 的配置

根据实验地址分配，PC2 的配置如图 12.3 所示。

图 12.3 PC2 的配置

3) PC3 的配置

根据实验地址分配，PC3 的配置如图 12.4 所示。

第 12 章 防火墙配置

图 12.4　PC3 的配置

2. 服务器的配置

1)　Web Server 配置

(1) Web Server 基本配置。

根据实验地址分配，Web Server 的基本配置如图 12.5 所示。

图 12.5　Web　Server 基本配置

(2) Web Server 功能配置。

在 D 盘上创建 web 文件夹，用来存放要发布的网页，在 Web Server 上选择"服务器信息标签"，单击单选按钮"HttpServer"，在配置标签中选择 D:\web，单击"启动"按钮完成 Web Server 功能配置。Web Server 功能配置如图 12.6 所示。

图 12.6　Web　Server 功能配置

2) FTP Server 配置

(1) FTP Server 基本配置。

根据实验地址分配，FTP Server 的基本配置如图 12.7 所示。

图 12.7　FTP　Server 基本配置

(2) Web Server 功能配置。

在 D 盘上创建 ftp 文件夹，用来存放要文件，在 FTP Server 上选择"服务器信息"标签，单击单选按钮"FtpServer"，在配置标签中选择 D:\ftp，单击"启动"按钮完成 FTP Server 功能配置。FTP Server 功能配置如图 12.8 所示。

图 12.8　FTP　Server 功能配置

3. Client 的配置

1) Client1 的配置

根据实验地址分配，Client1 的配置如图 12.9 所示。

图 12.9　Client1 的配置

2) Client2 的配置

根据实验地址分配，Client2 的配置如图 12.10 所示。

图 12.10　Client2 的配置

4. FW1 配置

1) FW1 的启动

第一次启动 FW1，需要输入华为防火墙默认用户名 admin，默认口令 Admin@123。可以修改口令，按照提示，先后输入旧口令、新口令 Hbeutc123、确认新口令 Hbeutc123。

```
Username:admin          //默认用户名: admin
Password:               //默认口令: Admin@123
The password needs to be changed. Change now? [Y/N]: y
Please enter old password:      //输入旧口令: Admin@123
Please enter new password:      //新口令: Hbeutc123
Please confirm new password:    //确认新口令: Hbeutc123
Info: Your password has been changed. Save the change to survive a reboot.
**************************************************************
*       Copyright (C) 2014-2018 Huawei Technologies Co., Ltd.    *
*                     All rights reserved.                       *
*          Without the owner's prior written consent,            *
*      no decompiling or reverse-engineering shall be allowed.   *
**************************************************************
```

2) FW1 的基本配置

```
<Huawei>system-view
[Huawei]undo info-center enable
[Huawei]sysname FW1
[FW1]interface GigabitEthernet 1/0/0
[FW1-GigabitEthernet1/0/0]ip address 192.168.10.254 24
[FW1-GigabitEthernet1/0/0]service-manage enable
[FW1-GigabitEthernet1/0/0]service-manage ping permit
[FW1-GigabitEthernet1/0/0]interface GigabitEthernet 1/0/2
[FW1-GigabitEthernet1/0/2]ip address 192.168.20.254 24
[FW1-GigabitEthernet1/0/2]service-manage enable
[FW1-GigabitEthernet1/0/2]service-manage ping permit
[FW1-GigabitEthernet1/0/2]interface GigabitEthernet 1/0/1
[FW1-GigabitEthernet1/0/1]ip address 211.85.8.1 24
[FW1-GigabitEthernet1/0/1]service-manage enable
[FW1-GigabitEthernet1/0/1]service-manage ping permit
[FW1-GigabitEthernet1/0/1]quit
```

3) FW1 的路由配置

```
[FW1]ip route-static 0.0.0.0 0 211.85.8.2
```

4) FW1 的安全区域配置

```
[FW1]firewall zone trust
[FW1-zone-trust]add interface GigabitEthernet 1/0/2
[FW1-zone-trust]firewall zone dmz
[FW1-zone-dmz]add interface GigabitEthernet 1/0/0
[FW1-zone-dmz]firewall zone untrust
[FW1-zone-untrust]add interface GigabitEthernet 1/0/1
```

5) FW1 的安全策略配置

(1) 放行 Trust 区域到 DMZ 区域的 outbound 策略。

```
[FW1]security-policy
[FW1-policy-security]rule name t2d
[FW1-policy-security-rule-t2d]source-zone trust
[FW1-policy-security-rule-t2d]destination-zone dmz
[FW1-policy-security-rule-t2d]action permit
[FW1-policy-security-rule-t2d]quit
```

(2) 放行 Trust 区域到 Untrust 区域的 outbound 策略。

```
[FW1]security-policy
[FW1-policy-security]rule name t2u
[FW1-policy-security-rule-t2u]source-zone trust
[FW1-policy-security-rule-t2u]destination-zone untrust
[FW1-policy-security-rule-t2u]action permit
[FW1-policy-security-rule-t2u]quit
```

(3) 放行 Untrust 区域到 Local 区域的 inbound 策略。

```
[FW1]security-policy
[FW1-policy-security]rule name u2l
[FW1-policy-security-rule-u2l]source-zone untrust
[FW1-policy-security-rule-u2l]destination-zone local
[FW1-policy-security-rule-u2l]action permit
[FW1-policy-security-rule-u2l]quit
```

(4) 放行 Untrust 区域到 Trust 区域的 inbound 策略。

```
[FW1]security-policy
[FW1-policy-security]rule name u2t
[FW1-policy-security-rule-u2t]source-zone untrust
[FW1-policy-security-rule-u2t]destination-zone trust
[FW1-policy-security-rule-u2t]action permit
[FW1-policy-security-rule-u2t]quit
```

(5) 允许分公司访问 DMZ 区域的 inbound 策略。

```
[FW1]security-policy
[FW1-policy-security]rule name u2d
[FW1-policy-security-rule-u2z]rule name u2d
[FW1-policy-security-rule-u2d]source-address 211.85.9.0 24
[FW1-policy-security-rule-u2d]destination-zone dmz
[FW1-policy-security-rule-u2d]action permit
[FW1-policy-security-rule-u2d]quit
```

6) FW1 的 NAT 配置

```
[FW1]nat address-group 1
[FW1-address-group-1]section 0 211.85.8.101 211.85.8.200
[FW1-address-group-1]mode pat
[FW1-address-group-1]quit
```

7) FW1 的安全区域配置

```
[FW1]nat server 0 protocol tcp global GigabitEthernet 1/0/1 2323 inside 192.168.10.3 telnet
[FW1]nat server 1 protocol tcp global 211.85.8.108 www inside 192.168.10.1 www
[FW1]nat server 2 protocol tcp global 211.85.8.121 ftp inside 192.168.10.2 ftp
```

5. R0 的配置

1) R0 的基本配置

```
<Huawei>system-view
[Huawei]undo info-center enable
```

```
[Huawei]sysname R0
[R0]interface GigabitEthernet 0/0/0
[R0-GigabitEthernet0/0/0]ip address 211.85.16.254 24
[R0-GigabitEthernet0/0/0]interface GigabitEthernet 0/0/1
[R0-GigabitEthernet0/0/1]ip address 211.85.9.1 24
[R0-GigabitEthernet0/0/1]interface GigabitEthernet 0/0/2
[R0-GigabitEthernet0/0/2]ip address 211.85.8.2 24
[R0-GigabitEthernet0/0/2]quit
```

2) R0 的路由配置

```
[R0]ip route-static 192.168.40.0 24 211.85.9.2
[R0]ip route-static 192.168.10.0 24 211.85.8.1
[R0]ip route-static 192.168.20.0 24 211.85.8.1
```

6. R1 的配置

1) R1 的基本配置

```
<Huawei>system-view
[Huawei]undo info-center enable
[Huawei]sysname R1
[R1]interface GigabitEthernet 0/0/0
[R1-GigabitEthernet0/0/0]ip address 211.85.9.2 24
[R1-GigabitEthernet0/0/0]interface GigabitEthernet 0/0/1
[R1-GigabitEthernet0/0/1]ip address 192.168.40.254 24
[R1-GigabitEthernet0/0/1]quit
```

2) R1 的路由配置

```
[R1]ip route-static 0.0.0.0 0 211.85.9.1
```

3) R1 的 NAT 配置

在 R1 上配置 NAT，定义基本 ACL 2023，ACL 规则源地址 192.168.40.0/24，直接使用 GE 0/0/0 接口的公有 IP 地址作为 NAT 转换后的 IP 地址。

```
[R1]acl 2023
[R1-acl-basic-2023]rule 5 permit source 192.168.40.0 0.0.0.255
[R1-acl-basic-2023]interface GigabitEthernet 0/0/0
[R1-GigabitEthernet0/0/0]nat outbound 2023
```

第 12 章 防火墙配置

7. R2 的配置

1) R2 的基本配置

```
<Huawei>system-view
[Huawei]sysname R2
[R2]interface GigabitEthernet 0/0/0
[R2-GigabitEthernet0/0/0]ip address 192.168.10.3 24
[R2-GigabitEthernet0/0/0]quit
```

2) R2 的路由配置

```
[R2]ip route-static 0.0.0.0 0 192.168.10.254
```

3) R2 的 Telnet 服务配置

在 R2 上配置 Telnet 服务，设置使用密码验证登录方式，密码为 Hbeutc123。

```
[R2]user-interface vty 0 4
[R2-ui-vty0-4]authentication-mode password
Please configure the login password (maximum length 16):Hbeutc123
[R2-ui-vty0-4]
```

8. R3 的配置

1) R3 的基本配置

```
<Huawei>system-view
[Huawei]sysname R3
[R3]interface GigabitEthernet 0/0/0
[R3-GigabitEthernet0/0/0]ip address 192.168.40.2 24
[R3-GigabitEthernet0/0/0]quit
```

2) R3 的路由配置

```
[R3]ip route-static 0.0.0.0 0 192.168.40.254
```

9. R4 的配置

1) R4 的基本配置

```
<Huawei>system-view
[Huawei]sysname R4
[R4]interface GigabitEthernet 0/0/0
[R4-GigabitEthernet0/0/0]ip address 211.85.16.2 24
[R4-GigabitEthernet0/0/0]quit
```

2) R4 的路由配置

```
[R4]ip route-static 0.0.0.0 0 211.85.16.254
```

10. 测试

1) Untrust 区域中不同 Telnet 客户端访问 Telnet Server

(1) 在 R3 上访问 Telnet Server。

```
<R3>telnet 211.85.8.1 2323
 Press CTRL_] to quit telnet mode
 Trying 211.85.8.1 ...
 Connected to 211.85.8.1 ...
Login authentication
Password:    //输入Hbeutc123
<R2>
```

结果显示，R3 正常访问 Telnet Server，分公司网络可访问 Telnet Server。

(2) 在 R4 上访问 Telnet Server。

```
<R4>telnet 211.85.8.1 2323
 Press CTRL_] to quit telnet mode
 Trying 211.85.8.1 ...
 Error: Can't connect to the remote host
```

结果显示，Untrust 区域非分公司网络无法访问 Telnet Server。

2) Untrust 区域中不同客户端访问 FTP Server

(1) 在 R3 上访问 FTP Server。

在 R3 上访问 FTP Server，结果如图 12.11 所示。

第 12 章 防火墙配置

图 12.11　R3 正常访问 FTP　Server

(2) 在 R4 上访问 FTP Server。

在 R4 上访问 FTP Server，结果如图 12.12 所示。

图 12.12　R4 无法访问 FTP　Server

3) Untrust 区域中不同客户端访问 Web Server

(1) 在 R3 上访问 Web Server.

在 R3 上访问 Web Server，结果如图 12.13 所示。

图 12.13　R3 正常访问 Web　Server

(2) 在 R4 上访问 Web Server。

在 R4 上访问 Web Server，结果如图 12.14 所示。

图 12.14　R4 正常访问 Web　Server

4) 不同安全区域之间测试

(1) Untrust 区域访问 DMZ 区域。

在 Client1 上测试与 Web Server 的连通性，结果如图 12.15 所示。

第 12 章 防火墙配置

图 12.15　Untrust 区域无法访问 DMZ 区域

(2) Untrust 区域访问 Trust 区域。

在 Client1 上测试与 PC1 的连通性，结果如图 12.16 所示。

图 12.16　Untrust 区域可以访问 Trust 区域

(3) Trust 区域访问 DMZ 区域。

在 PC1 上使用 ping 命令测试与 Web Server 的连通性。

```
PC>ping 192.168.10.1
Ping 192.168.10.1: 32 data bytes, Press Ctrl_C to break
From 192.168.10.1: bytes=32 seq=1 ttl=254 time=62 ms
From 192.168.10.1: bytes=32 seq=2 ttl=254 time=32 ms
From 192.168.10.1: bytes=32 seq=3 ttl=254 time=31 ms
From 192.168.10.1: bytes=32 seq=4 ttl=254 time=31 ms
From 192.168.10.1: bytes=32 seq=5 ttl=254 time=31 ms
--- 192.168.10.1 ping statistics ---
 5 packet(s) transmitted
 5 packet(s) received
 0.00% packet loss
 round-trip min/avg/max = 31/37/62 ms
```

结果显示，Trust 区域可以访问 DMZ 区域。

(4) Trust 区域访问 Untrust 区域。

在 PC1 上使用 ping 命令测试与 Client1 的连通性。

```
PC>ping 211.85.16.1
Ping 211.85.16.1: 32 data bytes, Press Ctrl_C to break
From 211.85.16.1: bytes=32 seq=1 ttl=253 time=47 ms
```

```
From 211.85.16.1: bytes=32 seq=2 ttl=253 time=47 ms
From 211.85.16.1: bytes=32 seq=3 ttl=253 time=47 ms
From 211.85.16.1: bytes=32 seq=4 ttl=253 time=63 ms
From 211.85.16.1: bytes=32 seq=5 ttl=253 time=62 ms
--- 211.85.16.1 ping statistics ---
 5 packet(s) transmitted
 5 packet(s) received
 0.00% packet loss
 round-trip min/avg/max = 47/53/63 ms
```

结果显示，Trust 区域可以访问 Untrust 区域。

(5) DMZ 区域访问 Untrust 区域。

在 R2 上使用 ping 命令测试与 Client1 的连通性。

```
<R2>ping -c 1 211.85.16.1
  PING 211.85.16.1: 56  data bytes, press CTRL_C to break
    Request time out
  --- 211.85.16.1 ping statistics ---
    1 packet(s) transmitted
    0 packet(s) received
    100.00% packet loss
```

结果显示，DMZ 区域无法访问 Untrust 区域。

(6) DMZ 区域访问 Trust 区域。

在 R2 上使用 ping 命令测试与 PC1 的连通性。

```
<R2>ping -c 1 192.168.20.1
  PING 192.168.20.1: 56  data bytes, press CTRL_C to break
    Request time out
  --- 192.168.20.1 ping statistics ---
    1 packet(s) transmitted
    0 packet(s) received
    100.00% packet loss
```

结果显示，DMZ 区域无法访问 Trust 区域。

第 13 章 双机热备份配置

13.1 路由器热备份配置

一、原理概述

虚拟路由冗余协议(virtual router redundancy protocol, VRRP)是由 IETF 提出的解决局域网中配置静态网关出现单点失效现象的路由协议。VRRP 广泛应用在边界网络中，它允许主机使用双路由器，在实际第一跳路由器使用失败的情形下仍能够维护路由器间的连通性。就是两个网关，一个主网关，一个备份网关，主网关失效，备份网关继续转发数据，主网关恢复正常，便自动切换为主网关转发数据。

VRRP 使用选举机制来确定路由器的状态（Master 或 Backup）。运行 VRRP 的一组路由器对外组成一个虚拟路由器，其中一台路由器处于 Master 状态，其他路由器处于 Backup 状态。

运行 VRRP 的路由器都会发送和接收 VRRP 通告消息，在通告消息中包含了自身的 VRRP 优先级信息。VRRP 通过比较路由器的优先级进行选举，优先级高的路由器将成为主路由器，其他路由器都为备份路由器。

虚拟路由器和 VRRP 路由器都有自己的 IP 地址（虚拟路由器的 IP 地址可以和 VRRP 备份组内的某个路由器的接口地址相同）。如果 VRRP 组中存在 IP 地址拥有者，即虚拟地址与某台 VRRP 路由器的地址相同时，则 IP 地址拥有者将成为主路由器，并且拥有最高优先级 255。如果 VRRP 组中不存在 IP 地址拥有者，则 VRRP 路由器将通过比较优先级来确定主路由器。路由器可配置的优先级范围为 1～254，默认情况下 VRRP 路由器的优先级为 100。当优先级相同时，VRRP 将通过比较 IP 地址来进行选举，IP 地址大的路由器将成为主路由器。

二、实验目的

1. 理解 VRRP 的应用场景

2. 理解 VRRP 虚拟路由器的配置
3. 掌握修改 VRRP 优先级的方法
4. 掌握查看 VRRP 主备状态的方法

三、实验内容

本实验模拟企业网络环境,某企业内部局域网通过网关路由器接入 ISP 路由器,为提高网络的可靠性,采用 VRRP 构成双网关路由器,实现冗余备份的同时,可实现负载均衡。

四、实验拓扑

路由器热备份配置拓扑如图 13.1 所示。

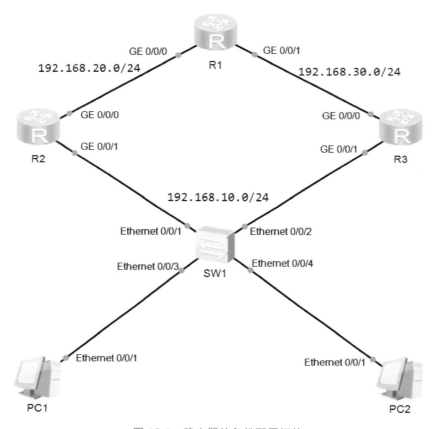

图 13.1　路由器热备份配置拓扑

五、实验地址分配

实验地址分配如表 13.1 所示。

第 13 章 双机热备份配置

表 13.1 实验地址分配

设备	接口	IP 地址	子网掩码	默认网关
R1（AR2220）	GE 0/0/0	192.168.20.1	255.255.255.0	N/A
	GE 0/0/1	192.168.30.1	255.255.255.0	N/A
	Loopback 0	2.2.2.2	255.255.255.255	
R2（AR2220）	GE 0/0/0	192.168.20.2	255.255.255.0	N/A
	GE 0/0/1	192.168.10.252	255.255.255.0	N/A
R3（AR2220）	GE 0/0/0	192.168.30.1	255.255.255.0	N/A
	GE 0/0/1	192.168.10.253	255.255.255.0	N/A
PC1	Ethernet 0/0/1	192.168.10.1	255.255.255.0	192.168.10.254
PC2	Ethernet 0/0/1	192.168.10.2	255.255.255.0	192.168.10.254

六、实验步骤

1. PC 的配置

根据实验地址分配表，PC1 的配置如图 13.2 所示。

图 13.2 PC1 的配置

根据实验地址分配表，PC2 的配置如图 13.3 所示。

图 13.3 PC2 的配置

2. R1 的配置

1）基本配置

根据实验地址分配表，在 R1 上配置基本信息。

```
<Huawei>system-view
[Huawei]undo info-center enable
[Huawei]sysname R1
[R1]interface GigabitEthernet 0/0/0
[R1-GigabitEthernet0/0/0]ip address 192.168.20.1 24
[R1-GigabitEthernet0/0/0]interface GigabitEthernet 0/0/1
[R1-GigabitEthernet0/0/1]ip address 192.168.30.1 24
[R1-GigabitEthernet0/0/1]interface loopback 0
[R1-LoopBack0]ip address 2.2.2.2 32
```

2) OSPF 配置

```
[R1]ospf 1
[R1-ospf-1]area 0
[R1-ospf-1-area-0.0.0.0]network 192.168.20.0 0.0.0.255
[R1-ospf-1-area-0.0.0.0]network 192.168.30.0 0.0.0.255
[R1-ospf-1-area-0.0.0.0]network 2.2.2.2 0.0.0.0
```

3. R2 的配置

1) 基本配置

根据实验地址分配表，在 R2 上配置基本信息。

```
<Huawei>system-view
[Huawei]undo info-center enable
[Huawei]sysname R2
[R2]interface GigabitEthernet 0/0/0
[R2-GigabitEthernet0/0/0]ip address 192.168.20.2 24
[R2-GigabitEthernet0/0/0]interface GigabitEthernet 0/0/1
[R2-GigabitEthernet0/0/1]ip address 192.168.10.252 24
[R2-GigabitEthernet0/0/1]quit
```

2) OSPF 配置

```
[R2]ospf 1
[R2-ospf-1]area 0
[R2-ospf-1-area-0.0.0.0]network 192.168.20.0 0.0.0.255
[R2-ospf-1-area-0.0.0.0]network 192.168.10.0 0.0.0.255
```

第 13 章 双机热备份配置

4. R3 的配置

(1) 基本配置

根据实验地址分配表，在 R3 上配置基本信息。

```
<Huawei>system-view
[Huawei]undo info-center enable
[Huawei]sysname R3
[R3]interface GigabitEthernet 0/0/0
[R3-GigabitEthernet0/0/0]ip address 192.168.30.1 24
[R3-GigabitEthernet0/0/0]interface GigabitEthernet 0/0/1
[R3-GigabitEthernet0/0/1]ip address 192.168.10.253 24
[R3-GigabitEthernet0/0/1]quit
```

2) OSPF 配置

```
[R3]ospf 1
[R3-ospf-1]area 0
[R3-ospf-1-area-0.0.0.0]network 192.168.30.0 0.0.0.255
[R3-ospf-1-area-0.0.0.0]network 192.168.10.0 0.0.0.255
```

5. VRRP 配置

在 R2 和 R3 上配置 VRRP 协议，使用 vrrp vrid 1 virtual-ip 命令创建 VRRP 备份组，指定 R1 与 R2 处于同一个 VRRP 备份组内，VRRP 备份组号为 1，配置虚拟 IP 为 192.168.10.254。虚拟 IP 地址必须和当前接口 IP 处于同一个网段。

```
[R2]interface GigabitEthernet 0/0/1
[R2-GigabitEthernet0/0/1]vrrp vrid 1 virtual-ip 192.168.10.254

[R3]interface GigabitEthernet 0/0/1
[R3-GigabitEthernet0/0/1]vrrp vrid 1 virtual-ip 192.168.10.254
```

经过配置后，PC 将使用虚拟路由器 IP 地址作为默认网关。

6. 测试

1) 查看 VRRP

分别在 R2 和 R3 上使用 display vrrp brief 命令查看 VRRP 配置结果。

```
[R2]display vrrp brief
Total:1    Master:1    Backup:0    Non-active:0
```

```
VRID  State      Interface         Type        Virtual IP
-----------------------------------------------------------------
1     Master     GE0/0/1           Normal      192.168.10.254
```

结果显示，R2 为主路由器。

```
[R3]display vrrp brief
Total:1     Master:0     Backup:1     Non-active:0
VRID  State      Interface         Type        Virtual IP
-----------------------------------------------------------------
1     Backup     GE0/0/1           Normal      192.168.10.254
```

结果显示，R3 为备份路由器。
在 PC1 上使用 tracert 命令测试到 R1 的路由信息。

```
PC>tracert 2.2.2.2
traceroute to 2.2.2.2, 8 hops max
(ICMP), press Ctrl+C to stop
 1  192.168.10.252    32 ms   46 ms   47 ms
 2  2.2.2.2           32 ms   46 ms   47 ms
```

结果显示，PC1 到 R1 是通过 R2（主路由器）转发的。

2) 修改 R3 的 VRRP 优先级

```
[R3]interface GigabitEthernet 0/0/1
[R3-GigabitEthernet0/0/1]vrrp vrid 1 priority 120
[R3-GigabitEthernet0/0/1]quit
```

分别在 R2 和 R3 上使用 display vrrp brief 命令查看 VRRP 配置结果。

```
[R2]display vrrp brief
Total:1     Master:0     Backup:1     Non-active:0
VRID  State      Interface         Type        Virtual IP
-----------------------------------------------------------------
1     Backup     GE0/0/1           Normal      192.168.10.254
```

结果显示，R2 为备份路由器。

```
[R3]display vrrp brief
Total:1    Master:1      Backup:0    Non-active:0
VRID  State       Interface              Type      Virtual IP
----------------------------------------------------------------
1     Master      GE0/0/1                Normal    192.168.10.254
```

结果显示，R3 为主路由器。

在 PC1 上使用 tracert 命令测试到 R1 的路由信息。

```
PC>tracert 2.2.2.2
traceroute to 2.2.2.2, 8 hops max
(ICMP), press Ctrl+C to stop
1  192.168.10.253    31 ms   47 ms   47 ms
2  2.2.2.2           62 ms   63 ms   62 ms
```

结果显示，PC1 到 R1 是通过 R3（主路由器）转发的。

13.2 交换机热备份配置

一、原理概述

虚拟路由冗余协议 VRRP 是一种用于提高网络可靠性的容错协议。通过 VRRP，可以在主机的下一跳设备出现故障时，及时将业务切换到备份设备，从而保障网络通信的连续性和可靠性。

多生成树协议 MSTP 不仅涉及多个 MST（生成树实例），而且可以划分为多个 MST 区域。通常，MSTP 网络可以包含一个或多个 MST 域，每个 MST 域可以包含一个或多个 MST。每个 MST 由运行 STP/RSTP/MSTP 的交换设备组成，这些交换设备通过 MSTP 计算后形成树状网络。

MSTP 与 VRRP 结合使用，可以实现负载分担和冗余备份，是一种基础的组网方式。

二、实验目的

1. 理解 VRRP+MSTP 进行交换机双机热备份的应用场景
2. 理解 VRRP 和 MSTP 的基本原理
3. 掌握 VRRP+MSTP 进行交换机双机热备份的配置过程

三、实验内容

本实验模拟企业网络环境,某企业建筑物理上分为行政楼和综合楼,为提高网络的可靠性,采用双汇聚+双核心的双归链路提高网络高可靠性(链路级及设备级的负载均衡及冗余备份),利用 MSTP(多生成树协议)+VRRP(虚拟路由冗余协议)提高可靠性,实现冗余备份的同时,可实现负载均衡,MSTP 中创建多个生成树实例,实现 VLAN 间负载均衡,不同 VLAN 的流量按照不同的路径转发。VRRP 中创建多个备份组,各备份组指定不同的 Master 与 Backup,实现虚拟路由的负载均衡。

四、实验拓扑

交换机热备份配置拓扑如图 13.4 所示。

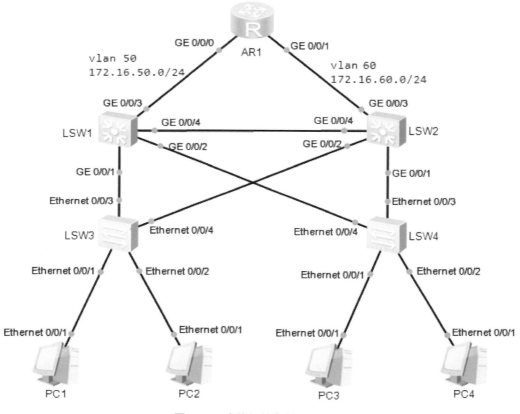

图 13.4 交换机热备份配置拓扑

五、实验地址分配

实验地址分配如表 13.2 所示。

第 13 章　双机热备份配置

表 13.2　实验地址分配

设备	接口	IP 地址	子网掩码	默认网关
AR1（AR2220）	GE 0/0/0	172.16.50.2	255.255.255.0	N/A
	GE 0/0/1	172.16.60.2	255.255.255.0	N/A
	Loopback 0	2.2.2.2	255.255.255.255	N/A
LSW1	VLANIF10	172.16.10.253	255.255.255.0	N/A
	VLANIF20	172.16.20.253	255.255.255.0	N/A
	VLANIF30	172.16.30.253	255.255.255.0	N/A
	VLANIF40	172.16.40.253	255.255.255.0	N/A
	VLANIF50	172.16.50.1	255.255.255.0	N/A
LSW2	VLANIF10	172.16.10.252	255.255.255.0	N/A
	VLANIF20	172.16.20.252	255.255.255.0	N/A
	VLANIF30	172.16.30.252	255.255.255.0	N/A
	VLANIF40	172.16.40.252	255.255.255.0	N/A
	VLANIF60	172.16.50.2	255.255.255.0	N/A
PC1	Ethernet 0/0/1	172.16.10.1	255.255.255.0	172.16.10.254
PC2	Ethernet 0/0/1	172.16.20.1	255.255.255.0	172.16.20.254
PC3	Ethernet 0/0/1	172.16.30.1	255.255.255.0	172.16.30.254
PC4	Ethernet 0/0/1	172.16.40.1	255.255.255.0	172.16.40.254

六、实验步骤

1. LSW1 的配置

1) 基本配置

在 LSW1 上使用 VLAN 的相关命令创建 vlan 10、vlan 20、vlan 30 和 vlan 40，并配置各活动接口为 Turnk 类型，允许所有 VLAN 通过。

```
<Huawei>system-view
[Huawei]sysname LSW1
[LSW1]undo info-center enable
[LSW1]vlan batch 10 20 30 40 50
[LSW1]interface GigabitEthernet 0/0/1
[LSW1-GigabitEthernet0/0/1]port link-type trunk
[LSW1-GigabitEthernet0/0/1]port trunk allow-pass vlan all
[LSW1-GigabitEthernet0/0/1]interface GigabitEthernet 0/0/2
[LSW1-GigabitEthernet0/0/2]port link-type trunk
[LSW1-GigabitEthernet0/0/2]port trunk allow-pass vlan all
[LSW1-GigabitEthernet0/0/2]interface GigabitEthernet 0/0/3
```

```
[LSW1-GigabitEthernet0/0/3]port link-type access
[LSW1-GigabitEthernet0/0/3]port default vlan 50
[LSW1-GigabitEthernet0/0/3]interface GigabitEthernet 0/0/4
[LSW1-GigabitEthernet0/0/4]port link-type trunk
[LSW1-GigabitEthernet0/0/4]port trunk allow-pass vlan all
[LSW1-GigabitEthernet0/0/4]interface vlanif 50
[LSW1-Vlanif50]ip address 172.16.50.1 24
```

2) VRRP 的配置

使用 VRRP+MSTP 组网做设备备份和流量负载均衡，PC1/2 走 LSW1，LSW1 作为 vlan 10 和 vlan 20 的主路由且为根桥，作为 vlan 30 和 vlan 40 的备路由，PC3/4 走 LSW2，LSW2 作为 vlan 30 和 vlan 40 主路由且为根桥，作为 vlan 10 和 vlan 20 的备路由，各交换机之间配置 Trunk。

在 LSW1 上，使用 vrrp 命令在 vlanif 10 接口上创建名为 vrid 1 虚拟组，虚拟 IP 地址为 172.16.10.254，优先级设为 120；在 vlanif 20 接口上创建名为 vrid 2 虚拟组，虚拟 IP 地址为 172.16.20.254，优先级设为 120；在 vlanif 30 接口上创建名为 vrid 3 虚拟组，虚拟 IP 地址为 172.16.30.254，优先级设为默认 100；在 vlanif 40 接口上创建名为 vrid 4 虚拟组，虚拟 IP 地址为 172.16.40.254，优先级设为默认 100。在 vlanif 10 和 vlanif 20 上使用 track 命令跟踪上行接口 GE 0/0/3，当上行接口发生故障时，降低优先级成为备用设备。

```
[LSW1]interface vlanif 10
[LSW1-Vlanif10]ip address 172.16.10.253 24
[LSW1-Vlanif10]vrrp vrid 1 virtual-ip 172.16.10.254
[LSW1-Vlanif10]vrrp vrid 1 priority 120
[LSW1-Vlanif10]vrrp vrid 1 track interface GigabitEthernet 0/0/3 reduce 30
[LSW1-Vlanif10]interface vlanif 20
[LSW1-Vlanif20]ip address 172.16.20.253 24
[LSW1-Vlanif20]vrrp vrid 2 virtual-ip 172.16.20.254
[LSW1-Vlanif20]vrrp vrid 2 priority 120
[LSW1-Vlanif20]vrrp vrid 2 track interface GigabitEthernet 0/0/3 reduced 30
[LSW1-Vlanif20]interface vlanif 30
[LSW1-Vlanif30]ip address 172.16.30.253 24
[LSW1-Vlanif30]vrrp vrid 3 virtual-ip 172.16.30.254
[LSW1-Vlanif30]interface vlanif 40
[LSW1-Vlanif40]ip address 172.16.40.253 24
[LSW1-Vlanif40]vrrp vrid 4 virtual-ip 172.16.40.254
[LSW1-Vlanif40]quit
```

3) MSTP 的配置

在 LSW1 上使用 stp enable 命令启动 STP，使用 stp mode mstp 命令将 STP 的模式设置为 MSTP，进入 STP 域配置视图，使用 region-name 命令配置域名为 vrrp，其他设备域名一致；实例 1 关联 vlan 10 和 vlan 20，实例 2 关联 vlan 30 和 vlan 40；退出域配置视图，在全局模式下，将 LSW1 设为实例 1 的根桥、设为实例 2 的次根桥。

第 13 章 双机热备份配置

```
[LSW1]stp enable
[LSW1]stp mode mstp
[LSW1]stp region-configuration
[LSW1-mst-region]region-name vrrp
[LSW1-mst-region]instance 1 vlan 10 20
[LSW1-mst-region]instance 2 vlan 30 40
[LSW1-mst-region]quit
[LSW1]stp instance 1 root primary
[LSW1]stp instance 2 root secondary
```

4) DHCP 的配置

在 LSW1 上，分别为 vlan 10、vlan 20、vlan 30、vlan 40 配置 DHCP。

```
[LSW1]ip pool vlan10
[LSW1-ip-pool-vlan10]network 172.16.10.0 mask 255.255.255.0
[LSW1-ip-pool-vlan10]gateway-list 172.16.10.254
[LSW1-ip-pool-vlan10]excluded-ip-address 172.16.10.252 172.16.10.253
[LSW1-ip-pool-vlan10]ip pool vlan20
[LSW1-ip-pool-vlan20]network 172.16.20.0 mask 255.255.255.0
[LSW1-ip-pool-vlan20]gateway-list 172.16.20.254
[LSW1-ip-pool-vlan20]excluded-ip-address 172.16.20.252 172.16.20.253
[LSW1-ip-pool-vlan20]ip pool vlan30
[LSW1-ip-pool-vlan30]network 172.16.30.0 mask 255.255.255.0
[LSW1-ip-pool-vlan30]gateway-list 172.16.30.254
[LSW1-ip-pool-vlan30]excluded-ip-address 172.16.30.252 172.16.30.253
[LSW1-ip-pool-vlan30]ip pool vlan40
[LSW1-ip-pool-vlan40]network 172.16.40.0 mask 255.255.255.0
[LSW1-ip-pool-vlan40]gateway-list 172.16.40.254
[LSW1-ip-pool-vlan40]excluded-ip-address 172.16.40.252 172.16.40.253
[LSW1-ip-pool-vlan40]quit
```

在 LSW1 上，分别在 vlanif 10、vlanif 20、vlanif 30 和 vlanif 40 接口中开启全局地址池 DHCP。

```
[LSW1]interface vlanif 10
[LSW1-Vlanif10]dhcp select global
[LSW1-Vlanif10]interface vlanif 20
[LSW1-Vlanif20]dhcp select global
[LSW1-Vlanif20]interface vlanif 30
[LSW1-Vlanif30]dhcp select global
[LSW1-Vlanif30]interface vlanif 40
[LSW1-Vlanif40]dhcp select global
```

5) OSPF 配置

```
[LSW1]ospf 1
[LSW1-ospf-1]area 0
[LSW1-ospf-1-area-0.0.0.0]network 172.16.10.0 0.0.0.255
[LSW1-ospf-1-area-0.0.0.0]network 172.16.20.0 0.0.0.255
[LSW1-ospf-1-area-0.0.0.0]network 172.16.30.0 0.0.0.255
[LSW1-ospf-1-area-0.0.0.0]network 172.16.40.0 0.0.0.255
[LSW1-ospf-1-area-0.0.0.0]network 172.16.50.0 0.0.0.255
```

2. LSW2 的配置

1) 基本配置

在 LSW2 上使用 VLAN 的相关命令创建 vlan 10、vlan 20、vlan 30 和 vlan 40，并配置各活动接口为 Turnk 类型，允许所有 vlan 通过。

```
<Huawei>system-view
[Huawei]undo info-center enable
[Huawei]sysname LSW2
[LSW2]vlan batch 10 20 30 40 60
[LSW2]interface GigabitEthernet 0/0/1
[LSW2-GigabitEthernet0/0/1]port link-type trunk
[LSW2-GigabitEthernet0/0/1]port trunk allow-pass vlan all
[LSW2-GigabitEthernet0/0/1]interface GigabitEthernet 0/0/2
[LSW2-GigabitEthernet0/0/2]port link-type trunk
[LSW2-GigabitEthernet0/0/2]port trunk allow-pass vlan all
[LSW2-GigabitEthernet0/0/2]interface GigabitEthernet 0/0/3
[LSW2-GigabitEthernet0/0/3]port link-type access
[LSW2-GigabitEthernet0/0/3]port default vlan 60
[LSW2-GigabitEthernet0/0/3]interface GigabitEthernet 0/0/4
[LSW2-GigabitEthernet0/0/4]port link-type trunk
[LSW2-GigabitEthernet0/0/4]port trunk allow-pass vlan all
[LSW2-GigabitEthernet0/0/4]interface vlanif 60
[LSW1-Vlanif60]ip address 172.16.60.1 24
```

2) VRRP 的配置

使用 VRRP+MSTP 组网做设备备份和流量负载均衡，PC1/2 走 LSW1，LSW1 作为 vlan 10 和 vlan 20 的主路由且为根桥，作为 vlan 30 和 vlan 40 的备路由，PC3/4 走 LSW2，LSW2 作为 vlan 30 和 vlan 40 主路由且为根桥，作为 vlan 10 和 vlan 20 的备路由，各交换机之间配置 Trunk。

```
[LSW2]interface vlanif 10
[LSW2-Vlanif10]ip address 172.16.20.252 24
[LSW2-Vlanif10]vrrp vrid 1 virtual-ip 172.16.10.254
```

第 13 章 双机热备份配置

```
[LSW2-Vlanif10]vrrp vrid 1 priority 100
[LSW2-Vlanif10]interface vlanif 20
[LSW2-Vlanif20]ip address 172.16.20.252 24
[LSW2-Vlanif20]vrrp vrid 2 virtual-ip 172.16.20.254
[LSW2-Vlanif20]interface vlanif 30
[LSW2-Vlanif30]ip address 172.16.30.252 24
[LSW2-Vlanif30]vrrp vrid 3 virtual-ip 172.16.30.254
[LSW2-Vlanif30]vrrp vrid 3 priority 120
[LSW2-Vlanif30]vrrp vrid 3 track interface GigabitEthernet 0/0/3 reduce 30
[LSW2-Vlanif30]interface vlanif 40
[LSW2-Vlanif40]ip address 172.16.40.252 24
[LSW2-Vlanif40]vrrp vrid 4 virtual-ip 172.16.40.254
[LSW2-Vlanif40]vrrp vrid 4 priority 120
[LSW2-Vlanif40]vrrp vrid 4 track interface GigabitEthernet 0/0/3 reduce 30
[LSW2-Vlanif40]quit
```

3) MSTP 的配置

在 LSW2 上使用 stp enable 命令启动 STP，使用 stp mode mstp 命令将 STP 的模式设置为 MSTP，进入 STP 域配置视图，使用 region-name 命令配置域名为 vrrp（与 LSW1 一致）；实例 1 关联 vlan 10 和 vlan 20，实例 2 关联 vlan 30 和 vlan 40；退出域配置视图，在全局模式下，将 LSW2 设为实例 2 的根桥、设为实例 1 的次根桥。

```
[LSW2]stp enable
[LSW2]stp mode mstp
[LSW2]stp region-configuration
[LSW2-mst-region]region-name vrrp
[LSW2-mst-region]instance 1 vlan 10 20
[LSW2-mst-region]instance 2 vlan 30 40
[LSW2-mst-region]quit
[LSW2]stp instance 2 root primary
[LSW2]stp instance 1 root secondary
```

4) DHCP 的配置

在 LSW2 上，分别为 vlan 10、vlan 20、vlan 30、vlan 40 配置 DHCP。

```
[LSW2]ip pool vlan10
[LSW2-ip-pool-vlan10]network 172.16.10.0 mask 255.255.255.0
[LSW2-ip-pool-vlan10]gateway-list 172.16.10.254
[LSW2-ip-pool-vlan10]excluded-ip-address 172.16.10.252 172.16.10.253
[LSW2-ip-pool-vlan10]ip pool vlan20
[LSW2-ip-pool-vlan20]network 172.16.20.0 mask 255.255.255.0
[LSW2-ip-pool-vlan20]gateway-list 172.16.20.254
[LSW2-ip-pool-vlan20]excluded-ip-address 172.16.20.252 172.16.20.253
[LSW2-ip-pool-vlan20]ip pool vlan30
```

```
[LSW2-ip-pool-vlan30]network 172.16.30.0 mask 255.255.255.0
[LSW2-ip-pool-vlan30]gateway-list 172.16.30.254
[LSW2-ip-pool-vlan30]excluded-ip-address 172.16.30.252 172.16.30.253
[LSW2-ip-pool-vlan30]ip pool vlan40
[LSW2-ip-pool-vlan40]network 172.16.40.0 mask 255.255.255.0
[LSW2-ip-pool-vlan40]gateway-list 172.16.40.254
[LSW2-ip-pool-vlan40]excluded-ip-address 172.16.40.252 172.16.40.253
[LSW2-ip-pool-vlan40]quit
```

在 LSW2 上，分别在 vlanif 10、vlanif 20、vlanif 30 和 vlanif 40 接口中开启全局地址池 DHCP。

```
[LSW2]interface vlanif 10
[LSW2-Vlanif10]dhcp select global
[LSW2-Vlanif10]interface vlanif 20
[LSW2-Vlanif20]dhcp select global
[LSW2-Vlanif20]interface vlanif 30
[LSW2-Vlanif30]dhcp select global
[LSW2-Vlanif30]interface vlanif 40
[LSW2-Vlanif40]dhcp select global
```

5) OSPF 配置

```
[LSW2]ospf 1
[LSW2-ospf-1]area 0
[LSW2-ospf-1-area-0.0.0.0]network 172.16.10.0 0.0.0.255
[LSW2-ospf-1-area-0.0.0.0]network 172.16.20.0 0.0.0.255
[LSW2-ospf-1-area-0.0.0.0]network 172.16.30.0 0.0.0.255
[LSW2-ospf-1-area-0.0.0.0]network 172.16.40.0 0.0.0.255
[LSW2-ospf-1-area-0.0.0.0]network 172.16.60.0 0.0.0.255
```

3. AR1 的配置

1) 基本配置

根据实验地址分配表，为路由器 AR1 各接口配置 IP 地址。

```
<Huawei>system-view
[Huawei]undo info-center enable
[Huawei]sysname AR1
[AR1]interface GigabitEthernet 0/0/0
[AR1-GigabitEthernet0/0/0]ip address 172.16.50.2 24
[AR1-GigabitEthernet0/0/0]interface GigabitEthernet 0/0/1
[AR1-GigabitEthernet0/0/1]ip address 172.16.60.2 24
```

第 13 章 双机热备份配置

```
[AR1-GigabitEthernet0/0/1]interface LoopBack 0
[AR1-LoopBack0]ip address 2.2.2.2 32
```

2) OSPF 配置

```
[AR1]ospf 1
[AR1-ospf-1]area 0
[AR1-ospf-1-area-0.0.0.0]network 172.16.50.0 0.0.0.255
[AR1-ospf-1-area-0.0.0.0]network 172.16.60.0 0.0.0.255
[AR1-ospf-1-area-0.0.0.0]network 2.2.2.2 0.0.0.0
```

4. LSW3 的配置

在 LSW3 上，将 Ethernet 0/0/1 和 Ethernet 0/0/2 配置为 Access 模式，分别属于 vlan 10 和 vlan 20。将 Ethernet 0/0/3 和 Ethernet 0/0/4 配置为 Trunk 模式，允许所有 vlan 通过。

```
<Huawei>system-view
[Huawei]undo info-center enable
[Huawei]sysname LSW3
[LSW3]interface Ethernet 0/0/1
[LSW3-Ethernet0/0/1]port link-type access
[LSW3-Ethernet0/0/1]port default vlan 10
[LSW3-Ethernet0/0/1]interface Ethernet 0/0/2
[LSW3-Ethernet0/0/2]port link-type access
[LSW3-Ethernet0/0/2]port default vlan 20
[LSW3-Ethernet0/0/2]interface Ethernet 0/0/3
[LSW3-Ethernet0/0/3]port link-type trunk
[LSW3-Ethernet0/0/3]port trunk allow-pass vlan all
[LSW3-Ethernet0/0/3]interface Ethernet 0/0/4
[LSW3-Ethernet0/0/4]port link-type trunk
[LSW3-Ethernet0/0/4]port trunk allow-pass vlan all
```

5. LSW4 的配置

在 LSW4 上，将 Ethernet 0/0/1 和 Ethernet 0/0/2 配置为 Access 模式，分别属于 vlan 30 和 vlan 40。将 Ethernet 0/0/3 和 Ethernet 0/0/4 配置为 Trunk 模式，允许所有 vlan 通过。

```
<Huawei>system-view
[Huawei]undo info-center enable
[Huawei]sysname LSW4
[LSW4]interface Ethernet 0/0/1
[LSW4-Ethernet0/0/1]port link-type access
[LSW4-Ethernet0/0/1]port default vlan 30
```

```
[LSW4-Ethernet0/0/1]interface Ethernet 0/0/2
[LSW4-Ethernet0/0/2]port link-type access
[LSW4-Ethernet0/0/2]port default vlan 40
[LSW4-Ethernet0/0/2]interface Ethernet 0/0/3
[LSW4-Ethernet0/0/3]port link-type trunk
[LSW4-Ethernet0/0/3]port trunk allow-pass vlan all
[LSW4-Ethernet0/0/3]interface Ethernet 0/0/4
[LSW4-Ethernet0/0/4]port link-type trunk
[LSW4-Ethernet0/0/4]port trunk allow-pass vlan all
```

6. 测试

在 LSW1 上使用 display vrrp brief 命令查看 VRRP 配置信息。

```
[LSW1]display vrrp brief
VRID  State       Interface       Type      Virtual IP
--------------------------------------------------------
1     Master      Vlanif10        Normal    172.16.10.254
2     Master      Vlanif20        Normal    172.16.20.254
3     Backup      Vlanif30        Normal    172.16.30.254
4     Backup      Vlanif40        Normal    172.16.40.254
--------------------------------------------------------
Total:4    Master:2    Backup:2    Non-active:0
```

结果显示，LSW1 上的 vlanif 10 和 vlanif 20 的 VRRP 状态为 Master，vlanif 30 和 vlanif 40 的 VRRP 状态为 Backup，各接口对应的虚拟 IP 地址正确。

```
[LSW2]display vrrp brief
VRID  State       Interface       Type      Virtual IP
--------------------------------------------------------
1     Backup      Vlanif10        Normal    172.16.10.254
2     Backup      Vlanif20        Normal    172.16.20.254
3     Master      Vlanif30        Normal    172.16.30.254
4     Master      Vlanif40        Normal    172.16.40.254
--------------------------------------------------------
Total:4    Master:2    Backup:2    Non-active:0
```

结果显示，LSW2 上的 vlanif 30 和 vlanif 40 的 VRRP 状态为 Master，vlanif 10 和 vlanif 20 的 VRRP 状态为 Backup，各接口对应的虚拟 IP 地址正确。

另外在 PC1 上使用 ping -t 命令测试与 PC4 的连通性，任意断开 SW3 与 SW1 的链路，SW1 与 SW2 的链路，SW1 与 SW4 的链路，观察到 PC1 与 PC4 之间都能正常通信。

13.3 防火墙热备份配置

一、原理概述

作为 USG 防火墙最重要的功能之一，双机热备份极大地提高了设备的可靠性，当主用设备发生故障时，备用设备可以立即接管受影响的业务，从而显著减少业务中断的持续时间。

当防火墙上多个区域需要提供双机备份功能时，需要在一台防火墙上配置多个 VRRP 备份组。由于 USG 防火墙是状态防火墙，它要求报文的来回路径通过同一台防火墙。为了满足这个限制条件，就要求在同一台防火墙上的所有 VRRP 备份组状态保持一致，即需要保证在主防火墙上所有 VRRP 备份组都是主状态，这样所有报文都将从此防火墙上通过，而另外一台防火墙则充当备份设备。防火墙只支持两台设备做 VRRP，不像路由器支持多台设备做 VRRP。

VRRP 组管理协议（VRRP group management protocol，VGMP）提出 VRRP 管理组的概念，将同一台防火墙上的多个 VRRP 备份组都加到一个 VRRP 管理组，由管理组统一管理所有 VRRP 备份组。通过统一控制各 VRRP 备份组状态的切换，来保证管理组内的所有 VRRP 备份组状态都是一致的。

华为冗余协议（Huawei redundancy protocol，HRP）报文实际上是一个 VGMP 报文，承载在 VGMP 报文的 Data 域，HRP 的作用主要是确保主备设备配置信息一致，用于主用设备出现问题时，备用设备能快速切换至主用设备，同时保证了主备之间的数据同步。

二、实验目的

1. 理解 VRRP+VGMP 进行防火墙双机热备份的应用场景
2. 理解 VRRP、VGMP 和 HRP 的基本原理
3. 掌握 VRRP+VGMP 进行防火墙双机热备份的配置过程

三、实验内容

本实验模拟企业网络环境，某企业通过防火墙将网络分为 Trust 区域和 Untrust 区域，为提高网络的可靠性，利用 VRRP（虚拟路由冗余协议）+VGMP（VRRP 组管理协议）提高可靠性，实现冗余备份。VRRP 实现防火墙间的双机备份，VGMP 负责统一管理多个 VRRP 组，确保管理组内的 VRRP 备份组状态一致。

四、实验拓扑

防火墙热备份配置拓扑如图 13.5 所示。（注意：防火墙的 GE 0/0/0 接口为管理接口，不使用）

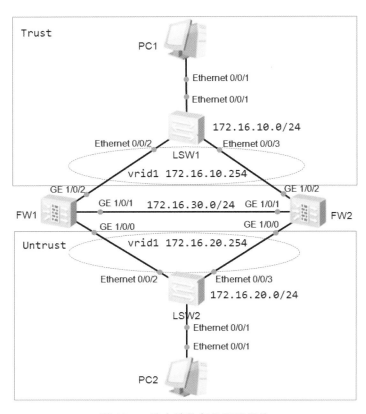

图 13.5　防火墙热备份配置拓扑

五、实验地址分配

实验地址分配如表 13.3 所示。

表 13.3　实验地址分配

设备	接口	IP 地址	子网掩码	默认网关
FW1 （USG6000V）	GE 1/0/2	172.16.10.100	255.255.255.0	N/A
	GE 1/0/0	172.16.20.100	255.255.255.0	N/A
	GE 1/0/1	172.16.30.100	255.255.255.0	N/A
FW2 （USG6000V）	GE 1/0/2	172.16.10.200	255.255.255.0	N/A
	GE 1/0/0	172.16.20.200	255.255.255.0	N/A
	GE 1/0/1	172.16.30.200	255.255.255.0	N/A
PC1	Ethernet 0/0/1	172.16.10.1	255.255.255.0	172.16.10.254
PC2	Ethernet 0/0/1	172.16.20.1	255.255.255.0	172.16.20.254

第 13 章 双机热备份配置

六、实验步骤

1. PC 的基本配置

PC1 的配置如图 13.6 所示。

图 13.6　PC1 的配置

PC2 的配置如图 13.7 所示。

图 13.7　PC2 的配置

2. FW1 的配置

1) FW1 的基本配置

首次登录防火墙时，按照提示仅将登录密码更改为 Hbeutc123，根据实验 IP 地址分配表，配置各接口的 IP 地址信息。

```
<USG6000V1>system-view
[USG6000V1]undo info-center enable
[USG6000V1]sysname FW1
[FW1]interface GigabitEthernet 1/0/2
[FW1-GigabitEthernet1/0/2]ip address 172.16.10.100 24
[FW1-GigabitEthernet1/0/2]interface GigabitEthernet 1/0/0
[FW1-GigabitEthernet1/0/0]ip address 172.16.20.100 24
[FW1-GigabitEthernet1/0/0]interface GigabitEthernet 1/0/1
[FW1-GigabitEthernet1/0/1]ip address 172.16.30.100 24
[FW1-GigabitEthernet1/0/1]quit
```

2) 防火墙的安全区域配置

使用 firewall zone trust 命令进入 Trust 安全区域，使用 add 命令将 GE 1/0/2 接口加入 Trust

安全区域。使用 firewall zone untrust 命令进入 Untrust 安全区域，使用 add 命令将 GE 1/0/0 接口加入 Untrust 安全区域，使用 firewall zone name hrp_zone 命令创建一个名为 hrp_zone 安全区域，用 set priority 20 命令设定安全级别为 20，使用 add 命令将 GE 1/0/1 接口加入 hrp_zone 安全区域。

```
[FW1]firewall zone trust
[FW1-zone-trust]add interface GigabitEthernet 1/0/2
[FW1-zone-trust]firewall zone untrust
[FW1-zone-untrust]add interface GigabitEthernet 1/0/0
[FW1-zone-untrust]quit
[FW1]firewall zone name hrp_zone
[FW1-zone-hrp_zone]set priority 20
[FW1-zone-hrp_zone]add interface GigabitEthernet 1/0/1
[FW1-zone-hrp_zone]quit
```

3) VRRP 配置

使用 vrrp 命令分别在 GE 1/0/2 上创建名为 vrid 1 虚拟组，虚拟 IP 地址 172.16.10.254，在 GE 1/0/0 上创建名为 vrid 2 虚拟组，虚拟 IP 地址 172.16.20.254。同一个虚拟组，虚拟 IP 地址必须一致。

```
[FW1]interface GigabitEthernet 1/0/2
[FW1-GigabitEthernet1/0/2]vrrp vrid 1 virtual 172.16.10.254 active
[FW1-GigabitEthernet1/0/2]service-manage enable
[FW1-GigabitEthernet1/0/2]service-manage ping permit
[FW1-GigabitEthernet1/0/2]interface GigabitEthernet 1/0/0
[FW1-GigabitEthernet1/0/0]service-manage enable
[FW1-GigabitEthernet1/0/0]service-manage ping permit
[FW1-GigabitEthernet1/0/0]vrrp vrid 2 virtual 172.16.20.254 active
[FW1-GigabitEthernet1/0/0]quit
```

3. FW2 的配置

1) FW2 的基本配置

首次登录防火墙时，按照提示仅将登录密码更改为 Hbeutc123，根据实验 IP 地址分配表，配置各接口的 IP 地址信息。

```
<USG6000V1>system-view
[USG6000V1]undo info-center enable
[USG6000V1]sysname FW2
[FW2]interface GigabitEthernet 1/0/2
[FW2-GigabitEthernet1/0/2]ip address 172.16.10.200 24
[FW2-GigabitEthernet1/0/2]interface GigabitEthernet 1/0/0
```

第 13 章 双机热备份配置

```
[FW2-GigabitEthernet1/0/0]ip address 172.16.20.200 24
[FW2-GigabitEthernet1/0/0]interface GigabitEthernet 1/0/1
[FW2-GigabitEthernet1/0/1]ip address 172.16.30.200 24
[FW2-GigabitEthernet1/0/1]quit
```

2) 防火墙的安全区域配置

使用 firewall zone trust 命令进入 Trust 安全区域，使用 add 命令将 GE 1/0/2 接口加入 Trust 安全区域。使用 firewall zone untrust 命令进入 Untrust 安全区域，使用 add 命令将 GE 1/0/0 接口加入 Untrust 安全区域，使用 firewall zone name hrp_zone 命令创建一个名为 hrp_zone 安全区域，用 set priority 20 命令设定安全级别为 20，使用 add 命令将 GE 1/0/1 接口加入 hrp_zone 安全区域。

```
[FW2]firewall zone trust
[FW2-zone-trust]add interface GigabitEthernet 1/0/2
[FW2-zone-trust]firewall zone untrust
[FW2-zone-untrust]add interface GigabitEthernet 1/0/0
[FW2-zone-untrust]quit
[FW2]firewall zone name hrp_zone
[FW2-zone-hrp_zone]set priority 20
[FW2-zone-hrp_zone]add interface GigabitEthernet 1/0/1
[FW2-zone-hrp_zone]quit
```

3) VRRP 配置

使用 vrrp 命令在 GE 1/0/2 上创建名为 vrid 1 虚拟组，虚拟 IP 地址为 172.16.10.254，在 GE 1/0/0 接口创建名为 vrid 2 的虚拟组，虚拟 IP 地址为 172.16.20.254。

```
[FW2]interface GigabitEthernet 1/0/2
[FW2-GigabitEthernet1/0/2]service-manage enable
[FW2-GigabitEthernet1/0/2]service-manage ping permit
[FW2-GigabitEthernet1/0/2]vrrp vrid 1 virtual-ip 172.16.10.254 standby
[FW2-GigabitEthernet1/0/2]interface GigabitEthernet 1/0/0
[FW2-GigabitEthernet1/0/0]service-manage enable
[FW2-GigabitEthernet1/0/0]service-manage ping permit
[FW2-GigabitEthernet1/0/0]vrrp vrid 2 virtual-ip 172.16.20.254 standby
[FW2-GigabitEthernet1/0/0]quit
```

4. 心跳线的配置

使用 hrp 命令完成心跳线（防火墙之间的直连线）的配置，设置 FW1 的接口 GE 1/0/1 对应远端 FW2 的 GE 1/0/1 的 IP 地址为 172.16.30.200，远端 IP 地址为 172.16.30.200；使用 hrp enable 命令启用 hrp；设置完毕后，在防火墙名称前出现 HRP_M，结果显示 FW1 为备用防火墙。在后续操作中，将 FW2 设置为备用防火墙，FW1 成为主用防火墙。

```
[FW1]hrp interface GigabitEthernet 1/0/1 remote 172.16.30.200
[FW1]hrp enable
HRP_S[FW1]
HRP_M[FW1]              //当防火墙FW2成为备用防火墙，FW1成为主用防火墙（回车可见）
```

在 FW2 上设置 GE 1/0/1 对应远端 FW1 的 GE 1/0/1，远端 IP 地址为 172.16.30.100，启用 hrp。

```
[FW2]hrp interface GigabitEthernet 1/0/1 remote 172.16.30.100
[FW2]hrp enable
HRP_S[FW2]              //将FW2设置成备份防火墙，此时FW1成为主用防火墙
```

5. 安全策略的配置

1) FW1 安全策略配置

在 FW1 上配置安全策略，设置规则 t2u，允许 Trust 安全区域访问 Untrust 区域。

```
HRP_M[FW1]security-policy (+B)
HRP_M[FW1-policy-security]rule name t2u (+B)
HRP_M[FW1-policy-security-rule-t2u]source-zone trust (+B)
HRP_M[FW1-policy-security-rule-t2u]destination-zone untrust (+B)
HRP_M[FW1-policy-security-rule-t2u]action permit (+B)
```

2) FW2 上查看同步安全策略

在主用防火墙 FW1 上配置安全策略，备用防火墙 FW2 上会自动备份。在 FW2 上面使用 display current-configuration 命令查看同步的安全策略。

```
HRP_S[FW2]display current-configuration
……
security-policy
 rule name t2u
  source-zone trust
  destination-zone untrust
  source-address 172.16.10.0 mask 255.255.255.0
  service icmp
  action permit
……
```

也可以在 FW2 上面使用 display security-policy rule all 命令查看同步的安全策略。

第 13 章　双机热备份配置

```
HRP_S[FW2]display security-policy rule all
2023-06-15 18:59:42.650
Total:2
RULE ID    RULE NAME              STATE      ACTION     HITS
-----------------------------------------------------------------
1          u2t                    enable     permit     0
0          default                enable     deny       0
-----------------------------------------------------------------
```

6. 测试

1) 测试 PC1 与 PC2 之间的连通性

在 PC1 上使用 ping 命令测试与 PC2 之间的连通性，同时，可以使用抓包软件分析 ICMP 分组通过主用防火墙 FW1。

```
PC>ping 172.16.20.1
Ping 172.16.20.1: 32 data bytes, Press Ctrl_C to break
From 172.16.20.1: bytes=32 seq=1 ttl=127 time=47 ms
From 172.16.20.1: bytes=32 seq=2 ttl=127 time=62 ms
From 172.16.20.1: bytes=32 seq=3 ttl=127 time=63 ms
From 172.16.20.1: bytes=32 seq=4 ttl=127 time=62 ms
From 172.16.20.1: bytes=32 seq=5 ttl=127 time=63 ms
--- 172.16.20.1 ping statistics ---
  5 packet(s) transmitted
  5 packet(s) received
  0.00% packet loss
  round-trip min/avg/max = 47/59/63 ms
```

结果显示，PC1 能正常访问 PC2。

2) 关闭 FW1 的 GE 1/0/2 接口

在 FW1 上使用 shutdown 命令关闭接口 GE 1/0/2。

```
HRP_M[FW1]interface GigabitEthernet 1/0/2
HRP_M[FW2-GigabitEthernet1/0/2]shutdown
```

在 FW2 上使用 display vrrp brief 命令查看 VRRP 信息。

```
HRP_M[FW2]display vrrp brief
2023-06-15 19:09:35.010
Total:2     Master:2      Backup:0      Non-active:0
VRID    State          Interface              Type       Virtual IP
```

313

```
----------------------------------------------------------------
1      Master      GE1/0/2           Vgmp      172.16.10.254
2      Master      GE1/0/0           Vgmp      172.16.20.254
```

结果显示,备用防火墙 FW2 变成主用防火墙,使用的 VRRP 管理协议为 VGMP。
再次使用 ping 命令测试 PC1 与 PC2 之间的连通性。

```
PC>ping 172.16.20.1
Ping 172.16.20.1: 32 data bytes, Press Ctrl_C to break
From 172.16.20.1: bytes=32 seq=1 ttl=127 time=46 ms
From 172.16.20.1: bytes=32 seq=2 ttl=127 time=63 ms
From 172.16.20.1: bytes=32 seq=3 ttl=127 time=46 ms
From 172.16.20.1: bytes=32 seq=4 ttl=127 time=47 ms
From 172.16.20.1: bytes=32 seq=5 ttl=127 time=63 ms
--- 172.16.20.1 ping statistics ---
 5 packet(s) transmitted
 5 packet(s) received
 0.00% packet loss
 round-trip min/avg/max = 46/53/63 ms
```

结果显示,PC1 仍然能正常访问 PC2,通过抓包软件,可以看到 ICMP 分组通过 FW2。

第 14 章 无线局域网配置

一、原理概述

AC+AP 是一种无线局域网的解决方案，由无线控制器（AC）和接入点（AP）组成。AC 作为无线网络的核心设备，负责接入点的管理、监控和控制，包括对接入点的配置、监控和故障排除等；而 AP 则负责连接无线终端设备并提供无线网络服务。

AC+AP 解决方案可以提供更加稳定和可靠的无线网络服务，具有扩展性强、管理和维护简单、网络稳定性高、安全性高等优点。它可以支持数百甚至数千个接入点的管理和控制，使得无线网络的管理和维护更加集中和简化。

在技术实现方面，AC+AP 解决方案主要采用了无线控制器与接入点之间的集中管理模式，采用了基于网络控制协议（NCP）的无线控制器–接入点协议（CAPWAP）进行通信，并采用了一系列安全技术保障数据的安全传输。

AC+AP 是一种功能强大、灵活可扩展的无线网络解决方案，采用了一系列先进的技术实现，可以提供更加稳定和可靠的无线网络服务，并适用于各种规模的企业和机构。

二、实验目的

1. 理解 AC+AP 的应用场景
2. 掌握 AC 配置的方法
3. 理解 WLAN 的工作原理

三、实验内容

本实验模拟企业网络环境，交换机 LSW1 为企业核心交换机，交换机 LSW2 和 LSW3 将接入层 AP1 和 AP2 接入核心交换机。核心交换机 LSW1 通过路由器 AR1 接入外网。企业无线网络 AC 采用旁挂组网方式，AC 与 AP 处于同一个二层网络。

四、实验拓扑

无线局域网的配置拓扑如图 14.1 所示。

计算机网络实践教程

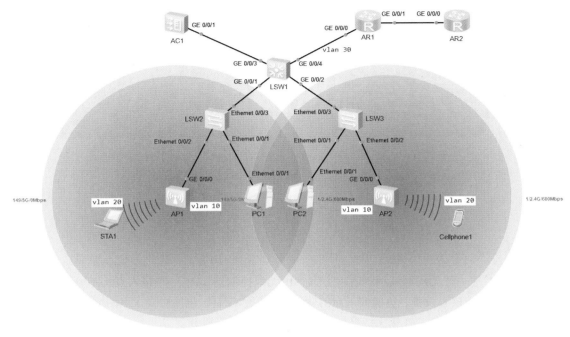

图 14.1 无线局域网配置拓扑

五、地址与参数设置

1. IP 地址分配

各设备接口 IP 地址和 VLAN 的分配如表 14.1 所示。

表 14.1 实验地址分配

设备名称	接口	VLAN	IP 地址
AR1	GE 0/0/0→LSW1:GE 0/0/4	30	192.168.30.2/24
	GE 0/0/1→AR2:GE 0/0/0		211.85.9.16/24
AR2	GE 0/0/0→AR1:GE 0/0/1		211.85.9.188/24
AC1	GE 0/0/1→LSW1:GE 0/0/3	Trunk	
	vlanif 10		192.168.10.254/24
LSW1	GE 0/0/1→LSW2:Ethernet 0/0/3	10	
	GE 0/0/2→LSW3:Ethernet 0/0/3	Trunk	
	GE 0/0/3→AC1:GE 0/0/1	Trunk	
	GE 0/0/4→AR1:GE 0/0/0	Trunk	
	vlanif 20		192.168.20.254/24
	vlanif 30		192.168.30.1/24

第 14 章　无线局域网配置

设备名称	接口	VLAN	IP 地址
LSW2	Ethernet 0/0/1→PC1: Ethernet 0/0/1	20	
	Ethernet 0/0/2→AP1:GE 0/0/0	10	
	Ethernet 0/0/3→LSW1:GE 0/0/1	Trunk	
LSW3	Ethernet 0/0/1→PC2: Ethernet 0/0/1	20	
	Ethernet 0/0/2→AP2:GE 0/0/0	10	
	Ethernet 0/0/3→LSW1:GE 0/0/2	Trunk	
AP1	GE 0/0/0→LSW2: Ethernet 0/0/2	10	
AP2	GE 0/0/0→LSW3: Ethernet 0/0/2	10	

2. AC 相关参数设置

AC 相关参数设置如表 14.2 所示。

表 14.2　AC 参数设置

配置项	配置参数
AP 管理 VLAN	vlan 10
STA 业务 VLAN	vlan 20
DHCP 服务器	AC 为 AP 分配 IP 地址：192.168.10.1～192.168.10.253
	LSW1 为 STA 分配 IP 地址：192.168.10.1～192.168.10.253
AP 组	名称：ap-group1
域管理模板	名称：default
	国家码：中国（CN）
CAPWAP	隧道源接口：vlanif10
	AP 认证模式：MAC 地址认证
安全模板	名称：hbeutc-wlan
	安全策略：WPA-WPA2+PSK+AES
	密码：hbeutc123
SSID 模板	名称：hbeutc-wlan
VAP 模板	名称：hbeutc-wlan
	转发模式：直接转发
	业务 VLAN：vlan 20

六、实验步骤

1. LSW1 的配置

1) 基本配置

根据实验地址分配表进行相应的基本配置，在 LSW1 上创建 vlan 10、vlan 20 和 vlan 30，在连接 AC 和 LSW2、LSW3 的接口上配置 Trunk，在连接 AR1 的接口上配置默认 vlan 30，并在 vlanif 30 接口上配置 IP 地址 192.168.30.1/24。

```
<Huawei>system-view
[Huawei]undo info-center enable
[Huawei]sysname LSW1
[LSW1]vlan batch 10 20 30
[LSW1]interface GigabitEthernet 0/0/1
[LSW1-GigabitEthernet0/0/1]port link-type trunk
[LSW1-GigabitEthernet0/0/1]port trunk allow-pass vlan 10 20
[LSW1-GigabitEthernet0/0/1]interface GigabitEthernet 0/0/2
[LSW1-GigabitEthernet0/0/2]port link-type trunk
[LSW1-GigabitEthernet0/0/2]port trunk allow-pass vlan 10 20
[LSW1-GigabitEthernet0/0/2]interface GigabitEthernet 0/0/3
[LSW1-GigabitEthernet0/0/3]port link-type trunk
[LSW1-GigabitEthernet0/0/3]port trunk allow-pass vlan 10 20
[LSW1-GigabitEthernet0/0/3]interface GigabitEthernet 0/0/4
[LSW1-GigabitEthernet0/0/4]port link-type access
[LSW1-GigabitEthernet0/0/4]port default vlan 30
[LSW1-GigabitEthernet0/0/4]interface vlanif 30
[LSW1-vlanif30]ip address 192.168.30.1 24
```

2) DHCP 配置

在 LSW1 上配置 DHCP 服务器，为终端 STA、PC 等分配 IP 地址，地址池为 192.168.20.0/24。接口 vlanif 20 的 IP 地址配置为 192.168.20.254/24，并在接口 vlanif 20 上选择全局 DHCP 分配模式。

```
[LSW1]dhcp enable
[LSW1]ip pool sta
[LSW1-ip-pool-sta]network 192.168.20.0 mask 24
[LSW1-ip-pool-sta]gateway-list 192.168.20.254
[LSW1-ip-pool-sta]dns-list 2.2.2.2
[LSW1-ip-pool-sta]quit
[LSW1]interface vlanif 20
[LSW1-vlanif20]ip address 192.168.20.254 24
[LSW1-vlanif20]dhcp select global
```

3) 路由配置

在 LSW1 上配置静态默认路由，使终端设备能够访问外网。

```
[LSW1]ip route-static 0.0.0.0 0 192.168.30.2
```

2. AC 的配置

1) 基本配置

在 AC1 上创建 vlan 10 和 vlan 20，在连接 LSW1 的接口上配置 Trunk。

```
<AC6605>system-view
[AC6605]undo info-center enable
[AC6605]sysname AC1
[AC1]vlan batch 10 20
[AC1]interface GigabitEthernet 0/0/1
[AC1-GigabitEthernet0/0/1]port link-type trunk
[AC1-GigabitEthernet0/0/1]port trunk allow-pass vlan 10 20
[AC1-GigabitEthernet0/0/1]quit
```

2) DHCP 配置

在 AC1 上配置 DHCP 服务器，为 AP 分配 IP 地址，地址池为 192.168.10.0/24。接口 vlanif 10 的 IP 地址配置为 192.168.10.254/24，并在接口 vlanif 10 上选择全局 DHCP 分配模式。

```
[AC1]dhcp enable
[AC1]ip pool ap
[AC1-ip-pool-ap]network 192.168.10.0 mask 255.255.255.0
[AC1-ip-pool-ap]gateway-list 192.168.10.254
[AC1-ip-pool-ap]interface vlanif 10
[AC1-vlanif10]ip address 192.168.10.254 24
[AC1-vlanif10]dhcp select global
```

3) WLAN 的配置

在 AC1 上使用 regulatory-domain-profile 命令创建名为 default 的域模板，使用 country-code 命令设置国家码为 cn，使用 ap-group name 命令创建名为 ap-group1 的 ap 组，使用 regulatory-domain-profile 命令在 ap-group1 组上应用 default 域模板。

```
[AC1]wlan
[AC1-wlan-view]regulatory-domain-profile name default
[AC1-wlan-regulate-domain-default]country-code cn
[AC1-wlan-regulate-domain-default]quit
```

```
[AC1-wlan-view]ap-group name ap-group1
[AC1-wlan-ap-group-ap-group1]regulatory-domain-profile default
Warning: Modifying the country code will clear channel,power and antenna gain
configurations of the radio and reset the AP.Continue?[Y/N]:y
[AC1-wlan-ap-group-ap-group1]quit
[AC1-wlan-view]quit
```

在 AC1 上使用 capwap source 命令配置 CAPWAP 隧道的源接口为 vlanif 10，在 wlan 视图中使用 ap auth-mode 命令配置 AP 认证模式为 MAC 地址认证，使用 ap-name 命令分别命名对应的 AP。最后将 AP 归入创建好的 ap-group1 组。

```
[AC1]capwap source interface vlanif 10
[AC1]wlan
[AC1-wlan-view]ap auth-mode mac-auth
[AC1-wlan-view]ap-id 0 ap-mac 00e0-fc76-6620
[AC1-wlan-ap-0]ap-name AP1
[AC1-wlan-ap-0]ap-group ap-group1
Warning: This operation may cause AP reset. If the country code changes, it
will clear channel, power and antenna gain configurations of the radio, Whether
to continue? [Y/N]:y
[AC1-wlan-ap-0]quit
[AC1-wlan-view]ap auth-mode mac-auth
[AC1-wlan-view]ap-id 1 ap-mac 00e0-fc70-63a0
[AC1-wlan-ap-1]ap-name AP2
[AC1-wlan-ap-1]ap-group ap-group1
Warning: This operation may cause AP reset. If the country code changes, it
will  clear channel, power and antenna gain configurations of the radio, Whether
to continue? [Y/N]:y
[AC1-wlan-ap-0]quit
```

在 AC1 的 wlan 视图下使用 security-profile name 命令创建名为 hbeutc-wlan 的安全模板，使用 security 命令设置安全策略为 wpa-wpa2，密码为 hbeutc123，加密方式为 psk+aes。在 AC1 的 wlan 视图下使用 ssid-profile name 命令创建名为 hbeutc-wlan 的 ssid 模板。

```
[AC1]wlan
[AC1-wlan-view]security-profile name hbeutc-wlan
[AC1-wlan-sec-prof-hbeutc-wlan]security wpa-wpa2 psk pass-phrase hbeutc123
aes
[AC1-wlan-sec-prof-hbeutc-wlan]quit
[AC1-wlan-view]ssid-profile name hbeutc-wlan
[AC1-wlan-ssid-prof-hbeutc-wlan]quit
```

在 AC1 的 wlan 视图下使用 **vap-profile** 命令创建名为 **hbeutc-wlan** 的 vap 模板，在 vap 模板中使用 **forward-mode** 设置转发方式为直接转发，使用 **service-vlan** 命令设置业务 vlan 为 vlan 20，使用 **security-profile** 应用安全模板 hbeutc-wlan，使用 **ssid-profile** 命令应用 ssid 模板 hbeutc-wlan，最后在 wlan 视图下进入 ap-group1 组，在 ap-group1 组模式下使用 **vap-profile** 命令绑定 vap 模板。

```
[AC1-wlan-view]vap-profile name hbeutc-wlan
[AC1-wlan-vap-prof-hbeutc-wlan]forward-mode direct-forward
[AC1-wlan-vap-prof-hbeutc-wlan]service-vlan vlan-id 20
[AC1-wlan-vap-prof-hbeutc-wlan]security-profile hbeutc-wlan
[AC1-wlan-vap-prof-hbeutc-wlan]ssid-profile hbeutc-wlan
[AC1-wlan-vap-prof-hbeutc-wlan]quit
[AC1-wlan-view]ap-group name ap-group1
[AC1-wlan-ap-group-ap-group1]vap-profile hbeutc-wlan wlan 1 radio all
```

4）查看登记的 AP 信息

```
[AC1]display ap all
Info: This operation may take a few seconds. Please wait for a moment.done.
Total AP information:
nor  : normal          [2]
--------------------------------------------------------------------------------
ID  MAC             Name  Group      IP              Type       State  STA Uptime
--------------------------------------------------------------------------------
0   00e0-fc76-6620  AP1   ap-group1  192.168.10.126  AP2050DN   nor    0   9S
1   00e0-fc70-63a0  AP2   ap-group1  192.168.10.175  AP2050DN   nor    0   6M:5S
--------------------------------------------------------------------------------
Total: 2
```

3. LSW2 的配置

在 LSW2 上创建 vlan 10 和 vlan 20，在连接 AP1 的接口上配置 Trunk，缺省 vlan 为 vlan 10，使 AP1 从 AC 上获取 IP 地址。在连接 PC1 的接口上配置默认 vlan 20，使 PC1 从 LSW1 上获取 IP 地址。

```
<Huawei>system-view
[Huawei]sysname LSW2
[LSW2]undo info-center enable
[LSW2]vlan batch 10 20
[LSW2]interface Ethernet 0/0/2
[LSW2-Ethernet0/0/2]port link-type trunk
[LSW2-Ethernet0/0/2]port trunk allow-pass vlan 10 20
[LSW2-Ethernet0/0/2]port trunk pvid vlan 10
```

```
[LSW2-Ethernet0/0/2]interface Ethernet 0/0/3
[LSW2-Ethernet0/0/3]port link-type trunk
[LSW2-Ethernet0/0/3]port trunk allow-pass vlan 10 20
[LSW2-Ethernet0/0/3]interface Ethernet 0/0/1
[LSW2-Ethernet0/0/1]port link-type access
[LSW2-Ethernet0/0/1]port default vlan 20
```

4. LSW3 的配置

在 LSW3 上创建 vlan 10 和 vlan 20，在连接 AP2 的接口上配置 Trunk，缺省 vlan 为 vlan 10，使 AP2 从 AC 上获取 IP 地址。在连接 PC2 的接口上配置默认 vlan 20，使 PC2 从 LSW1 上获取 IP 地址。

```
[Huawei]sysname LSW3
[LSW3]undo info-center enable
[LSW3]vlan batch 10 20
[LSW3]interface Ethernet 0/0/2
[LSW3-Ethernet0/0/2]port link-type trunk
[LSW3-Ethernet0/0/2]port trunk allow-pass vlan 10 20
[LSW3-Ethernet0/0/2]port trunk pvid vlan 10
[LSW3-Ethernet0/0/2]interface Ethernet 0/0/3
[LSW3-Ethernet0/0/3]port link-type trunk
[LSW3-Ethernet0/0/3]port trunk allow-pass vlan 10 20
[LSW3-Ethernet0/0/3]interface Ethernet 0/0/1
[LSW3-Ethernet0/0/1]port link-type access
[LSW3-Ethernet0/0/1]port default vlan 20
```

5. AR1 的配置

1) 基本配置

根据实验地址分配表进行相应的基本配置，在 AR1 上完成接口 IP 地址的配置。

```
[Huawei]sysname AR1
[AR1]interface GigabitEthernet 0/0/0
[AR1-GigabitEthernet0/0/0]ip address 192.168.30.2 24
[AR1]interface GigabitEthernet 0/0/1
[AR1-GigabitEthernet0/0/1]ip address 211.85.9.16 24
```

2) 路由配置

在 AR1 上配置向内和向外的静态路由信息，根据内部网络 IP 地址分配特点，选取聚合路由 192.168.0.0/16，外部网络路由选择默认路由。

第 14 章 无线局域网配置

```
[AR1]ip route-static 192.168.0.0 16 192.168.30.1
[AR1]ip route-static 0.0.0.0 0 211.85.9.188
```

3) NAT 配置

为了方便内网主机能够访问外网资源，在 AR1 上配置 NAT，设置基本 ACL，匹配内部所有网段。在连接 AR2 接口上使用 nat outbound 命令将内网地址映射为该端口的 IP 地址。

```
[AR1]acl 2000
[AR1-acl-basic-2000]rule permit source any
[AR1-acl-basic-2000]quit
[AR1]interface GigabitEthernet 0/0/1
[AR1-GigabitEthernet0/0/1]nat outbound 2000
```

6．AR2 的配置

1) 基本配置

根据实验地址分配表进行相应的基本配置，在 AR2 上完成接口 IP 地址的配置。

```
[Huawei]sysname AR2
[AR2]interface GigabitEthernet 0/0/0
[AR2-GigabitEthernet0/0/0]ip address 211.85.9.188 24
[AR2-GigabitEthernet0/0/0]quit
```

2) 路由配置

在 AR2 上配置向内的返程路由信息。

```
[AR2]ip route-static 0.0.0.0 0 211.85.9.16
```

7．测试

1) 测试 STA 获取 IP 地址

在 STA1 上使用 ipconfig 命令测试自动获取 IP 地址的情况。

```
STA>ipconfig
IPv4 address......................: 192.168.20.253
Subnet mask.......................: 255.255.255.0
Gateway...........................: 192.168.20.254
```

```
Physical address..................: 54-89-98-C2-23-8C
DNS server........................: 2.2.2.2
```

结果显示，STA1 能够正确获取 IP 地址信息。

在 cellphone1 上使用 **ipconfig** 命令测试自动获取 IP 地址的情况。同样地，可以测试 PC1 和 PC2 自动获取 IP 地址的情况。

```
STA>ipconfig
IPv4 address......................: 192.168.20.252
Subnet mask.......................: 255.255.255.0
Gateway...........................: 192.168.20.254
Physical address..................: 54-89-98-5F-49-D9
DNS server........................: 2.2.2.2
```

结果显示，cellphone1 能够正确获取 IP 地址信息。

2）测试连接在不同 AP 上的 STA 的连通性

在 STA1 上使用 **ping** 命令测试与 cellphone1 的连通性。同样地，可以测试 STA 与 PC 之间的连通性。

```
STA>ping 192.168.20.252
ping 192.168.20.252: 32 data bytes, Press Ctrl_C to break
From 192.168.20.252: bytes=32 seq=1 ttl=128 time=281 ms
From 192.168.20.252: bytes=32 seq=2 ttl=128 time=249 ms
From 192.168.20.252: bytes=32 seq=3 ttl=128 time=281 ms
From 192.168.20.252: bytes=32 seq=4 ttl=128 time=265 ms
From 192.168.20.252: bytes=32 seq=5 ttl=128 time=265 ms
--- 192.168.20.252 ping statistics ---
  5 packet(s) transmitted
  5 packet(s) received
  0.00% packet loss
  round-trip min/avg/max = 249/268/281 ms
```

结果显示，STA1 与 cellphone1 是连通的。

3）测试与外网的连通性

在 STA1 上使用 **ping** 命令测试与 AR2 的连通性。

```
STA>ping 211.85.9.188
ping 211.85.9.188: 32 data bytes, Press Ctrl_C to break
From 211.85.9.188: bytes=32 seq=2 ttl=253 time=219 ms
From 211.85.9.188: bytes=32 seq=2 ttl=253 time=217 ms
From 211.85.9.188: bytes=32 seq=3 ttl=253 time=188 ms
```

第 14 章 无线局域网配置

```
From 211.85.9.188: bytes=32 seq=4 ttl=253 time=156 ms
From 211.85.9.188: bytes=32 seq=5 ttl=253 time=172 ms
--- 211.85.9.188 ping statistics ---
 5 packet(s) transmitted
 4 packet(s) received
 20.00% packet loss
 round-trip min/avg/max = 0/183/219 ms
```

结果显示,STA1 与外网 AR2 是连通的。